Introducer

John Gribbin is the author of nearly 100 popular science books, including the bestselling *In Search of Schrödinger's Cat*. He has received awards for his writing both in the United States and in Britain. The holder of a PhD in astrophysics from the University of Cambridge, he still maintains links with research as a Visting Fellow in Astronomy at the University of Sussex, and is the leader of a team there that measured the age of the Universe. While still a student, he received the prestigious Annual Award of the Gravity Research Foundation in the United States, the only student, and the first Englishman working in England, ever to receive this award.

D0173432

ENCYCLOPÆDIA

THE **Britannica** GUIDE TO

THE 100 MOST INFLUENTIAL SCIENTISTS

The most important scientists from
Ancient Greece to the present day

Introduction by John Gribbin

ROBINSON

RUNNING PRESS
PHILADELPHIA · LONDON

Constable & Robinson Ltd
3 The Lanchesters
162 Fulham Palace Road
London W6 9ER
www.constablerobinson.com

Encyclopædia Britannica, Inc.
www.britannica.com

First published in the UK by Robinson,
an imprint of Constable & Robinson, 2008

A copy of the British Library Cataloguing in Publication
Data is available from the British Library

UK ISBN 978-1-84529-864-7

1 3 5 7 9 10 8 6 4 2

First published in the United States in 2008 by Running Press Book Publishers
All rights reserved under the Pan-American and International Copyright Conventions

US Library of Congress Control Number: 2007938551
US ISBN 978-0-7624-3421-3

Running Press Book Publishers
2300 Chestnut Street
Philadelphia, PA 19103-4371

www.runningpress.com

Printed and bound in the EU

CONTENTS

INTRODUCTION

Learning from the Lessons of History

John Gribbin

Science isn't what it used to be. Most scientists today work in large teams on projects which cost a lot of money and are inevitably, to a greater or lesser extent, steered by committees. Even theorists seldom work alone and they need computer time, which doesn't come cheap for the kind of supercomputers they use. One result of this is that projects have to have clear goals with widespread appeal before they are ever even started.

A century ago, the justification for building the 100-inch telescope that Edwin Hubble later used to discover the expansion of the universe was simply to find out more about the universe. In the late twentieth century, the main justification for building and launching the Hubble Space Telescope was determined in advance: the "Hubble Key Project" was formed to measure the expansion rate of the universe and thereby determine its age. By that time, the astronomers would never have got funding if they had simply asked for a big space telescope to explore the universe and find out what it's like. Similarly, the Large Hadron Collider (a particle accelerator) at

CERN, the European particle physics laboratory, is intended to "find the Higgs boson". Biologists were able to obtain funding to "map the human genome".

Of course, serendipitous discoveries come along the way; but there is little chance today for the kind of research carried out by most of the scientists whose lives and work are described in this book. So much has already been discovered. Isaac Newton could sit and wonder about so simple a thing as the fall of an apple from a tree, because nobody else had wondered about it in quite the same way before. A modern scientist intrigued by gravity has to learn all that Newton, Einstein and others discovered first, before attempting to extend the boundaries of this knowledge.

This is why the history of science, and especially a history which brings the scientists themselves alive for us, is so important. It shows us science as a more tangible human endeavour, that we can relate to more easily than we can to the search for the Higgs boson – but it also shows us how far we have come, since the speculations of those Greek philosophers who are now regarded as the first scientists. The unfolding story told in this particular history also gives the lie to one of the most pervasive, but misleading, myths about science – the idea that it proceeds in a series of revolutionary leaps.

A book like this makes it clear that science actually progresses in a series of relatively small steps, each one building on the work of earlier scientists. Isaac Newton famously wrote that if he had seen farther than others it was by "standing on the shoulders of Giants". There are several layers of meaning (not least that Newton intended it as an insult to his physically small but intellectually great contemporary Robert Hooke), but even taking it at face value it would be more appropriate to say that progress in science is made by "standing on the shoulders of midgets". The great breakthroughs came about principally from the culmination of years, often generations, of painstaking work. It is a long trek up a mountain path before a new vista is suddenly revealed in all its glory. The last step may be the most spectacular, but it would not have been possible without all the steps that went before.

Newton himself provides the perfect example of this. If anyone in the history of science might be thought of as a lone genius who single-handedly revolutionised our view of the world, it is surely Isaac Newton. And yet, we can trace a direct line to Newton's "discovery" of the law of universal gravitation in the 1680s from the work of William Gilbert, an Elizabethan physician who published a great book on magnetism in London in 1600. In his book, *De Magnete*, Gilbert set out clearly for the first time in print the essence of the scientific method of testing ideas by experiment, something that, for all their achievements, the philosophers of Ancient Greece had singularly failed to appreciate. Gilbert's book was read by, among others, Galileo Galilei, who spread the word and himself investigated, among other things, the nature of gravity, orbits, and the inertia of moving objects. By the 1680s, developing from the ideas of Galileo, Robert Hooke, Christopher Wren, and Edmond Halley, members of the newly established Royal Society, had got as far as speculating that gravity obeys an inverse square law, although they could not prove that this was the only possible explanation of the orbits of the planets around the Sun. It was Newton who put all of the pieces together, added his own insights, and came up with his idea of a universal inverse square law of gravitation, complete with a mathematical proof of its importance for orbits. But where would he have been without the others? And I haven't even mentioned the person usually cited, correctly, as a profound influence on Newton – Johannes Kepler, who discovered the laws of planetary motion!

The stories of people such as Gilbert, Galileo, and the others mentioned here highlight another fascinating feature of historical biography – the reminder that great discoveries were, and are, made by real people who had their daily lives to lead and experienced both the joys and tribulations of their times in much the same way as many of their contemporaries. Even at a professional level, there was often much more to them than the work for which they are remembered. Christopher Wren, for example, is best known today as an architect, but he was also a pioneering astronomer; conversely, Hooke is remembered as a scientist, but he was also an architect – Wren's partner in the reconstruction of London after the Great Fire of 1666 (the Monument to the fire is Hooke's work, as are several of the "Wren" churches). We also know that the early Fellows of the Royal Society used to meet socially with their friends in the fashionable coffee shops that sprang up in Restoration London, and, of course, they suffered the tribulations of both the plague of 1665 and the fire of 1666.

Scientists are not white-coated robots who have no lives outside the laboratory. The importance of spelling this out was brought home to me some time ago, when I began writing science fiction stories in collaboration with the novelist Douglas Orgill – now, alas, no longer with us. At first, my role was chiefly to ensure that the science was realistic, even if it was fictional. Sometimes, Douglas would call me, to ask how a character would react in the particular circumstances that the plot required. He was asking because he assumed that a scientist would react differently than a "normal" person would. Almost always, I would reply, "Douglas, how would *you* react?" He would tell me, and I would say, "well, fine, that's how the character would react." Soon, the penny dropped. Douglas began to treat his scientist characters as people with the same kind of loves, hates, and foibles as the other characters, and that particular kind of call stopped coming.

Where space permits, the backgrounds of the scientists and their work described here provide fascinating insights into "scientists as people" and of the social conditions that existed in the times they lived through. The lives and work of John Ray in the seventeenth century and of Joseph Priestley in the eighteenth century provide us with good examples of the continuing importance of religion and religious conflicts in Europe throughout most of the past six centuries, even if Ray and Priestley did not suffer quite so severely at the hands of the religious authorities as Giordano Bruno, burned at the stake for heresy, or Galileo, forced to renounce his scientific work.

As well as the interplay between science, scientists, and society, there was also the interplay between science and technology, which makes it particularly pleasing that the present volume does not take too narrow-minded a view of what "science" means, but includes great engineers, inventors, instrument-makers, and physicians. Throughout history, progress in science has led to progress in technology, and progress in technology has led to progress in science in a self-sustaining feedback. Better scientific instruments, such as telescopes and microscopes, led to the development of technologies which became the basis of better scientific instruments, such as electron microscopes and telescopes with digital imaging cameras.

It isn't as widely appreciated as it should be that one of the most important steps towards the quantum physics "revolution" in the twentieth century was the development in the nineteenth century of an efficient air pump, which made it possible to study the behaviour of things like beams of electrons in evacuated glass tubes. Without this seemingly mundane piece of equipment, J.J. Thomson would never have developed his understanding of the nature of the electron. One reason why the Ancient Greeks didn't understand electrons is that they didn't have good vacuum pumps, not that they weren't as clever as Thomson and his contemporaries.

Perhaps the neatest example of this feedback involves the growth of the science of thermodynamics and the practical implications leading to improvements in the steam engine during the nineteenth century. Starting with the story of James Watt, who pedants might claim was "merely" an instrument maker, the story can be traced through the work of people such as James Clerk Maxwell and Ludwig Boltzmann to become one of the most important features of our understanding of the physical world, encapsulated in the famous Second Law of Thermodynamics, which says, in a nutshell, "everything wears out". This is a more profound insight than it might seem at first, since it implies that the universe as we know it must have had a beginning a finite time ago, or it would have worn out already. Thus such a simple technological solution actually offers a direct link between James Watt's steam-engine and the Big Bang theory!

Another demonstration of the way science progresses incrementally rather than in a series of sudden leaps comes from the way in which new ideas have often occurred to different people at the same time: known as "multiple discoveries". The reason is not just that they are particularly clever ("but my goodness," as Lorelei Lee, the character played by Marilyn Monroe in the film *Gentlemen Prefer Blondes* commented in a different context, "doesn't it help?") but that previous discoveries have made the time ripe for the new insight.

The classic example of this is, of course, the theory of natural selection, which was hit upon independently by Charles Darwin and Alfred Russel Wallace, who both feature here, and announced their discovery in 1858. Incidentally, it's worth emphasising that the theory is indeed natural selection, not evolution. Evolution is a fact, observed in nature, and generations before Darwin and Wallace had been aware of the fact of evolution and puzzled over how to explain it – one of those earlier thinkers was Charles Darwin's own grandfather, Erasmus Darwin. Natural selection is the theory put forward to explain the fact of evolution. It is therefore not really a coincidence that two English naturalists of the nineteenth century should each be stimulated by their observations of the profusion of life in the tropics and the writings of Thomas Malthus to discover the process of natural selection, at the same time.

A slightly less familiar example is the discovery (or invention) of the periodic table of the elements, which is usually attributed to Dmitry Mendeleyev, who features here, but was actually a complicated story involving at least three of his contemporaries before something like the modern version of the periodic table emerged. Both the English industrial chemist John Newlands and the French mineralogist Alexandre Béguyer de Chancourtois realised independently that if the elements are arranged in order of their atomic weight, there is a repeating pattern of chemical properties.

Their ideas were published in the first half of the 1860s, when Béguyer was simply ignored while, in contrast, Newlands was savagely criticised for making such a ridiculous suggestion. In 1864, the German chemist and physician Lothar Meyer, unaware of any of this, published a hint at his own version in a textbook, and then developed a full account of the periodic table for a second edition, which did not appear in print until 1870. By which time Mendeleyev had presented his version of the periodic table to the world of chemistry, in complete ignorance of all the work along similar lines going on in England, France, and Germany.

Meyer always acknowledged Mendeleyev's priority, because Mendeleyev had been bold enough to publish his idea first, but Meyer and Mendeleyev shared the Davy medal of the Royal Society in 1882. Five years later, Newlands also received the Davy medal, leaving Béguyer as the odd man out.

Inevitably, in a project of this kind the choice of the hundred "most influential" scientists of all time must to some extent be subjective, and we can all make our own lists. Mine would certainly have included several people who do not make it here. Ibn al-Haytham, often referred to as Alhazen or Alhazan, was one of the great Arab experimental scientists, a pioneer in particular of optical studies, during the period around the year AD 1000 (he was born in 965 and died in 1040), when European science was standing still (if anything, regressing) in the centuries before the Renaissance. If it had not been for scientists like him and al-Khwārizmī, there would have been precious little knowledge with which to kick-start the Renaissance.

Charles Darwin's captain, Robert Fitzroy, did much more than act as the "driver" on the famous voyage of the HMS *Beagle*, and among other things invented weather forecasting, as the first Director of the UK Meteorological Office. Now, that really *was* influential! And coming right up to date, James Lovelock has surely been more influential than any living scientist, with his concept of the Earth as a living planet, Gaia.

The obvious question once you start listing the names of who to put in to a book like this is, who do you leave out to make room for them? And this raises the question of how scientists from different eras can be compared. It is impossible to say whether John Ray was more influential than Harry Kroto, or whether Ernest Rutherford was more influential than Thales of Miletus. Much of the appeal of the exercise surely lies in the broad sweep of history that is covered, with no attempt to say whether developments in one century are more or less significant in the long term than contributions in another period.

It is all too easy to look back from our present perspective and succumb to the temptation to see the twentieth century as a time of culmination in many areas of science. In the physical world, with the great theories of relativity and quantum physics, science has described the world on both the largest and the smallest scales, as well as everything in between; and in the living world with the understanding of the genetic code and the mechanism of evolution we seem to have the secret of life itself. But with the longer perspective, it is easy to see that we have been here before, even as recently as the end of the nineteenth century, when many scientists believed that all that remained was to dot a few i's and cross a few t's.

In a thousand years from now, perhaps our ideas about the nature of life and the universe will seem as strange as Johannes Kepler's "explanation" of the orbits of the planets in terms of nested polyhedra seems to us. And yet, although we are amused by this particular idea, we still acknowledge Kepler as one of the most influential scientists of all time. We can be sure that in a thousand years, whatever shape science may have taken, if twentieth-century science is remembered at all Albert Einstein, Francis Crick, and Richard Feynman will take their place in any equivalent survey.

The final, and perhaps most important, message from a survey of this kind is that science does indeed continue to progress, and that it is not always possible to predict the direction it will take. Perhaps the Large Hadron Collider will, indeed, find the Higgs boson, and the so-called Standard Model of particle physics will triumph once again. But perhaps it will not, and the Standard Model will have to be rebuilt. Contrary to what many people assume, it is the second possibility that excites the scientists. A career dotting i's and crossing t's is not one that appeals to many; what they long for is a new discovery that reveals new vistas at the top of the latest mountain path, with the prospect of a lifetime spent exploring the new opportunities that it opens up. It is that kind of opportunity, after all, that will give some lucky, and talented, people the opportunity to feature in a future edition of a book like this.

THALES OF MILETUS (*c.* 624–*c.* 546 BC)

One of the legendary Seven Wise Men,
or Sophoi, of antiquity, remembered primarily
for his cosmology.

No writings by Thales survive, and no contemporary sources exist, so his achievements are difficult to assess. The inclusion of his name in the canon of the legendary Seven Wise Men led to his idealization, and numerous acts and sayings, many of them no doubt spurious, were attributed to him. "Know thyself" and "Nothing in excess" are two examples.

Thales has been credited with the discovery of five geometric theorems: (1) that a circle is bisected by its diameter, (2) that in a triangle the angles opposite two sides of equal length are equal, (3) that opposite angles formed by intersecting straight lines are equal, (4) that the angle inscribed inside a semicircle is a right angle, and (5) that a triangle is determined if its base and the two angles at the base are given. His mathematical achievements are hard to determine, however, because of the practice at that time of crediting particular discoveries to men with a general reputation for wisdom.

The claim that Thales was the founder of European philosophy rests primarily on Aristotle (384–322 BC), who wrote that Thales was the first to suggest a single material substratum for the universe, namely water, or moisture. A likely consideration in this choice was the apparent motion that water exhibits, as seen in its ability to become vapour – for what changes or moves itself was thought by the Greeks to be close to life itself, and to Thales the entire universe was a living organism, nourished by exhalations from water. Thales' significance lies less in his choice of water as the essential substance than in his attempt to explain nature by the simplification of phenomena, as well as in his search for causes within nature itself rather than in the caprices of anthropomorphic gods. Like his successors the philosophers Anaximander (610–c. 546 BC) and Anaximenes of Miletus (flourished c. 545 BC), Thales is important in bridging the worlds of myth and reason.

PYTHAGORAS (c. 580–c. 500 BC)

Ancient Greek philosopher, mathematician, and founder of the Pythagorean brotherhood that contributed to the development of mathematics.

Pythagoras was born on Samos, an island in the Aegean Sea. He migrated to southern Italy in about 532 BC, apparently to escape Samos's tyrannical rule, and he established his ethico-political academy at Croton (now Crotone). He is generally credited with the theory of the functional significance of numbers in the objective world and in music. Other discoveries often attributed to him (e.g., the incommensurability of the side and diagonal of a square, and the Pythagorean theorem for right-angled triangles) were probably developed only later by the Pythagorean school. It is likely that the bulk of the intellectual tradition originating with Pythagoras himself belongs to mystical wisdom rather than to scientific scholarship.

The character of the original Pythagoreanism is controversial, and the conglomeration of disparate features that it displayed is confusing. Its fame rests, however, on some very influential ideas, not always correctly understood, that have been ascribed to it since antiquity. These ideas include those of (1) the metaphysic of number and the conception that reality, including music and astronomy, is, at its deepest level, mathematical in nature; (2) the use of philosophy as a means of spiritual purification; (3) the heavenly destiny of the soul and the possibility of its rising to union with the divine; (4) the appeal to certain symbols, sometimes mystical, such as the *tetraktys*, the golden section, and the harmony of the spheres; (5) the Pythagorean theorem, which holds that for a right-angled triangle a square drawn on the hypotenuse is equal in area to the sum of the squares drawn on its sides; and (6) the demand that members of the order shall observe a strict loyalty and secrecy.

HIPPOCRATES (*c.* 460–*c.* 375 BC)

Ancient Greek physician who lived during Greece's Classical period and is traditionally regarded as the father of medicine.

During his lifetime Hippocrates was admired as a physician and teacher. His younger contemporary Plato referred to him twice. In the *Protagoras* Plato called Hippocrates "the Asclepiad of Cos" who taught students for fees. It is now widely accepted that an "Asclepiad" was a physician belonging to a family that had produced well-known physicians for generations. Plato's second reference occurs in the *Phaedrus*, in which Hippocrates is referred to as a famous Asclepiad who had a philosophical approach to medicine. Meno, a pupil of Aristotle, specifically stated in his history of medicine the views of Hippocrates on the causation of diseases. These are the only surviving contemporary, or near-contemporary, references to Hippocrates.

His reputation, and myths about his life and his family, began to grow in the Hellenistic period, about a century after his death. During this period, the Museum of Alexandria in Egypt collected for its library literary material from preceding periods in celebration of the past greatness of Greece. It seems that the medical works that remained from the Classical period (among the earliest prose writings in Greek) were assembled as a group and called the *Corpus Hippocraticum* ("Works of Hippocrates"). Linguists and physicians subsequently wrote commentaries on them, and, as a result, all the virtues of the Classical medical works were eventually attributed to Hippocrates.

The merits of the Hippocratic writings are many, and, although they are of varying lengths and literary quality, they are all simple and direct, earnest in their desire to help, and lacking in technical jargon and elaborate argument. Prominent among these attractive works are the *Epidemics*, which give annual records of weather and associated diseases, along with individual case histories and records of treatment, collected from cities in northern Greece. Diagnosis and prognosis are frequent subjects. Other treatises explain how to set fractures and treat wounds, feed and comfort patients, and take care of the body to avoid illness. Treatises called *Diseases* deal with serious illnesses, proceeding from the head to the feet, giving symptoms, prognoses, and treatments. There are works on diseases of women, childbirth, and paediatrics. Prescribed medications, other than foods and local salves, are generally purgatives to rid the body of the noxious substances thought to cause disease. Some works argue that medicine is indeed a science, with firm principles and methods, although explicit medical theory is very rare; rather, the medicine depends on a mythology of how the body works and how its inner organs are connected. This myth is laboriously constructed from experience, but it must be remembered that there was neither systematic research nor dissection of human beings in Hippocrates' time. Hence, while much of the writing seems wise and correct, there are large areas where much is unknown.

PLATO (c. 428–c. 348 BC)

Ancient Greek philosopher who helped lay
the philosophical foundations of western culture.

Plato, the son of Ariston and Perictione, was born in Athens, or perhaps in Aegina, in about 428 BC, the year after the death of the great statesman Pericles. His family, on both sides, was among the most distinguished in Athens. Nothing is known about Plato's father's death, and it is assumed that he died when Plato was a boy. Perictione apparently married as her second husband her uncle Pyrilampes, a prominent supporter of Pericles, and Plato was probably brought up chiefly in his house.

The most important formative influence to which the young Plato was exposed was the philospher Socrates. It does not appear, however, that Plato belonged as a "disciple" to the circle of Socrates' intimates. Plato owed to Socrates his commitment to philosophy, his rational method, and his concern for ethical questions.

Plato's early ambitions – like those of most young men of his class – were probably political. A conservative faction urged him to enter public life under its auspices, but he wisely held back, and was soon repelled by its members' violent acts. After the fall of the oligarchy, he hoped for better things from the restored democracy. Eventually, however, he became convinced that there was no place for a man of conscience in Athenian politics. In 399 BC the Athenian democracy condemned Socrates to death, and Plato and other Socratic men took temporary refuge at Megara. His next few years are said to have been spent in extensive travels in Greece, Egypt, and Italy.

In about 387 BC Plato founded the Academy as an institute for the systematic pursuit of philosophical and scientific teaching and research. He presided over it for the rest of his life. Aristotle was a member of the Academy for 20 years, first as a student and then as a teacher. The Academy's interests encompassed a broad range of disciplines, including astronomy, biology, ethics, geometry, and rhetoric. All of the most important mathematical work of the fourth century BC was done by friends or students of Plato. The first students of conic sections, and possibly Theaetetus, the creator of solid geometry, were members of the Academy. Eudoxus of Cnidus – author of the doctrine of proportion expounded in Euclid's *Elements*, inventor of the method of finding the areas and volumes of curvilinear figures by exhaustion, and propounder of the astronomical scheme of concentric spheres adopted and altered by Aristotle – removed his school from Cyzicus to Athens for the purpose of cooperating with Plato. Archytas, the inventor of mechanical science, was a friend and correspondent of Plato. Nor were other sciences neglected. Speusippus, Plato's nephew and successor, was a voluminous writer on natural history, and Aristotle's biological works have been shown to belong largely to the early period in his career immediately after Plato's death. The comic poets found matter for mirth in the attention of the school to botanical classification. The Academy was particularly active in jurisprudence and practical legislation.

The Academy survived Plato's death. Though its interest in science waned and its philosophical orientation changed, it remained for two and a half centuries a focus of intellectual life. Its creation as a permanent society for the prosecution of both humane and exact sciences has been regarded – with pardonable exaggeration – as the first establishment of a university.

ARISTOTLE (384–322 BC)

Ancient Greek philosopher and scientist, one of the greatest intellectual figures of western history.

Aristotle was born on the Chalcidic peninsula of Macedonia, in northern Greece. His father, Nicomachus, was the physician of Amyntas III (reigned c. 393–c. 370 BC), king of Macedonia and grandfather of Alexander the Great (reigned 336–323 BC). Aristotle migrated to Athens after the death of his father in 367, and joined the Academy of Plato. He remained there for 20 years as Plato's pupil and colleague.

Many of Plato's later writings date from these decades, and they may reflect Aristotle's contributions to philosophical debate at the Academy. Some of Aristotle's writings also belong to this period, though mostly they survive only in fragments. Like his master, Aristotle wrote initially in dialogue form, and his early ideas show a strong Platonic influence.

During Aristotle's residence at the Academy, King Philip II of Macedonia (reigned 359–336 BC) waged war on a number of Greek city-states. The Athenians defended their independence only half-heartedly, and, after a series of humiliating concessions, they allowed Philip to become, by 338 BC, master of the Greek world. It cannot have been an easy time to be a Macedonian resident in Athens. Within the Academy, however, relations seem to have remained cordial. Aristotle always acknowledged a great debt to Plato; he took a large part of his philosophical agenda from Plato, and his teaching is more often a modification than a repudiation of Plato's doctrines.

When Plato died, in about 348 BC, his nephew Speusippus became head of the Academy and Aristotle left Athens. He migrated to Assus, a city on the north-western coast of Anatolia (in present-day Turkey), where Hermias, a graduate of the Academy, was ruler. Aristotle became a close friend of Hermias and eventually married his ward Pythias. Aristotle helped Hermias to negotiate an alliance with Macedonia, which angered the Persian king, who had Hermias treacherously arrested and put to death. Aristotle saluted Hermias's memory in *Ode to Virtue*, his only surviving poem.

While in Assus, and during the subsequent few years when he lived in the city of Mytilene on the island of Lesbos, Aristotle carried out extensive scientific research, particularly in zoology and marine biology. This work was summarized in a book later known, misleadingly, as *The History of Animals*, to which Aristotle added two short treatises, *On the Parts of Animals* and *On the Generation of Animals*. Although Aristotle did not claim to have founded the science of zoology, his detailed observations of a wide variety of organisms were quite without precedent. He – or one of his research assistants – must have been gifted with remarkably acute eyesight, since some of the features of insects that he accurately reports were not again observed until the invention of the microscope in the seventeenth century.

The scope of Aristotle's biological research is astonishing. Much of it is concerned with the classification of animals into genus and species: more than 500 species figure in his treatises, many of them described in detail. The myriad items of information about the anatomy, diet, habitat, modes of copulation, and reproductive systems of mammals, reptiles, fish, and insects are a mélange of minute investigation and vestiges of superstition. In some cases his unlikely stories about rare species of fish were proved accurate many centuries later. At other times he states clearly and fairly a biological problem that took millennia to solve, such as the nature of embryonic development.

Despite an admixture of the fabulous, Aristotle's biological works must be regarded as a stupendous achievement. His inquiries were conducted in a genuinely scientific spirit, and he was always ready to confess ignorance where evidence was insufficient. He insisted that whenever there is a conflict between theory and observation, one must trust observation, and that theories are to be trusted only if their results conform to the observed phenomena.

About eight years after the death of Hermias, in 343 or 342 BC, Aristotle was summoned by Philip II to the Macedonian capital at Pella to act as tutor to Philip's 13-year-old son, the future Alexander the Great. Little is known, however, of the content of Aristotle's instruction. By 326 BC Alexander had made himself master of an empire that stretched from the Danube to the Indus and included Libya and Egypt. Ancient sources report that during his campaigns Alexander arranged for biological specimens to be sent to his tutor from all parts of Greece and Asia Minor.

While Alexander was conquering Asia, Aristotle, now 50 years old, was in Athens. Just outside the city boundary, he established his own school in a gymnasium known as the Lyceum. He built a substantial library and gathered around him a group of brilliant research students, called "peripatetics" from the name of the cloister (*peripatos*) in which they walked and held their discussions. The Lyceum was not a private club like Plato's Academy; many of the lectures there were open to the general public and given free of charge.

Most of Aristotle's surviving works, with the exception of the zoological treatises, probably belong to this second Athenian sojourn. There is no certainty about their chronological order, and indeed it is probable that the main treatises – on physics, metaphysics, psychology, ethics, and politics – were constantly rewritten and updated. Every proposition of Aristotle is fertile in its ideas and full of energy, although his prose is generally neither lucid nor elegant.

Aristotle's works, though not as polished as Plato's, are systematic in a way that Plato's never were. Plato's dialogues shift constantly from one topic to another, always (from a modern perspective) crossing the boundaries between different philosophical or scientific disciplines. Indeed, there was no such thing as a specific intellectual discipline until Aristotle invented the notion during his Lyceum period.

Aristotle divided the sciences into three kinds: productive, practical, and theoretical. The productive sciences, naturally enough, are those that have a product. They include not only engineering and architecture, which have products such as bridges and houses, but also disciplines such as strategy and rhetoric, where the product is something less concrete, for example victory on the battlefield or in the courts. The practical sciences, most notably ethics and politics, are those that guide behaviour. The theoretical sciences are those that have no product and no practical goal but in which information and understanding are sought for their own sake. Aristotle divided the theoretical sciences into three groups: physics, mathematics, and theology. Physics as he understood it was equivalent to what would now be called "natural philosophy", or the study of nature: in this sense it encompasses not only the modern field of physics but also biology, chemistry, geology, psychology, and even meteorology.

In works such as *On Generation and Corruption* and *On the Heavens*, Aristotle presented a world-picture that included many features inherited from his pre-Socratic predecessors. From Empedocles (*c.* 490–430 BC) he adopted the view that the universe is ultimately composed of different combinations of the four fundamental elements of earth, water, air, and fire. Each element is characterized by the possession of a unique pair of the four elementary qualities of heat, cold, wetness, and dryness: earth is cold and dry, water is cold and wet, air is hot and wet, and fire is hot and dry. Each element has a natural place in an ordered cosmos, and each has an innate tendency to move toward this natural place. Thus, earthy solids naturally fall, while fire, unless prevented, rises ever higher. Aristotle's vision of the cosmos also owes much to Plato's dialogue *Timaeus.* As in that work, the Earth is at the centre of the universe, and around it the moon, the sun, and the other planets revolve in a succession of concentric crystalline spheres. The heavenly bodies are not compounds of the four terrestrial elements, but are made up of a superior fifth element, or "quintessence". In addition, the heavenly bodies have souls, or supernatural intellects, which guide them in their travels through the cosmos. The abiding value of his treatises on the physical sciences lies not in their particular scientific assertions but in their philosophical analyses of some of the concepts that pervade the physics of subsequent eras – concepts such as place, time, causation, and determinism.

When Alexander died in 323 BC, democratic Athens became uncomfortable for Macedonians. Aristotle, therefore, fled to Chalcis, where he died the following year. His will, which survives, made thoughtful provision for a large number of friends and dependents. To Theophrastus (*c.* 372–*c.* 287 BC), his successor as head of the Lyceum, he left his library, including his own writings, which were vast. Aristotle's surviving works amount to about one million words, though they probably represent only about a fifth of his total output.

By any reckoning, Aristotle's achievement is stupendous. His intellectual range was vast, covering most of the sciences and many of the arts, including biology, botany, chemistry, ethics, history, logic, metaphysics, rhetoric, philosophy of mind, philosophy of science, physics, poetics, political theory, psychology, and zoology. He was the founder of formal logic, devising for it a finished system that for centuries was regarded as the sum of the discipline. His writings in ethics and political theory as well as in metaphysics and the philosophy of science continue to be studied, and his work remains a powerful current in contemporary philosophical debate. Aristotle was the first genuine scientist in history, the first author whose surviving works contain detailed and extensive observations of natural phenomena, and the first philosopher to achieve a sound grasp of the relationship between observation and theory in scientific method. His Lyceum was the first research institute in which a number of scholars and investigators joined in collaborative inquiry and documentation. Finally, and not least important, he was the first person in history to build up a research library, a systematic collection of works to be used by his colleagues and to be bequeathed to posterity. Not only every philosopher but also every scientist is in his debt. He deserves the title Dante (1265–1321) gave him: "the master of those who know".

EUCLID (FLOURISHED c. 300 BC)

The most prominent mathematician of Greco-Roman antiquity, best known for his treatise on geometry, the *Elements*.

Euclid was born in Alexandria, Egypt. Of his life nothing is known except what the Greek philosopher Proclus (c. AD 410–

485) reports in his "summary" of famous Greek mathematicians. According to him, Euclid taught at Alexandria in the time of Ptolemy I Soter, who reigned over Egypt from 323 to 285 BC. Proclus supported his attribution of these dates to Euclid by writing "Ptolemy once asked Euclid if there was not a shorter road to geometry than through the Elements, and Euclid replied that there was no royal road to geometry."

Euclid compiled his *Elements* from a number of works of earlier men. Among these are Hippocrates of Chios (flourished *c.* 460 BC). The latest compiler before Euclid was Theudius, whose textbook was used in Plato's Academy and was probably the one used by Aristotle. The older elements were at once superseded by Euclid's and then forgotten. For his subject matter Euclid doubtless drew upon all his predecessors, but it is clear that the whole design of his work was his own, culminating in the construction of the five regular solids (pyramid, cube, octahedron, dodecahedron, icosahedron), now known as the Platonic solids.

A brief survey of the *Elements* belies a common belief that it concerns only geometry. This misconception may be caused by reading no further than Books I through IV, which cover elementary plane geometry. Euclid understood that building a logical and rigorous geometry (and mathematics) depends on the foundation – a foundation that Euclid began in Book I with 23 definitions (such as "a point is that which has no part" and "a line is a length without breadth"), five unproven assumptions that Euclid called postulates (now known as axioms), and five further unproven assumptions that he called common notions. Book I then proves elementary theorems about triangles and parallelograms, and ends with the Pythagorean theorem.

The subject of Book II has been called geometric algebra because it states algebraic identities as theorems about equivalent geometric figures. The book contains a construction of "the section", the division of a line into two parts such that the ratio of the larger to the smaller segment is equal to the ratio of the original line to the larger segment. (This division was renamed "the golden section" in the Renaissance after artists and architects rediscovered its pleasing proportions.) Book II also generalizes the Pythagorean theorem to arbitrary triangles, a result that is equivalent to the law of cosines. Book III deals with properties of circles, and Book IV with the construction of regular polygons, in particular the pentagon.

Book V shifts from plane geometry to expound a general theory of ratios and proportions that is attributed by Proclus (along with Book XII) to Eudoxus of Cnidus (*c.* 390–350 BC). While Book V can be read independently of the rest of the *Elements*, its solution to the problem of incommensurables (irrational numbers) is essential to later books. In addition, it formed the foundation for a geometric theory of numbers until an analytic theory developed in the late nineteenth century. Book VI applies this theory of ratios to plane geometry, mainly triangles and parallelograms, culminating in the "application of areas", a procedure for solving quadratic problems by geometric means.

Books VII–IX contain elements of number theory, where number (*arithmos*) means positive integers greater than 1. Beginning with 22 new definitions – such as unity, even, odd, and prime – these books develop various properties of the positive integers. For instance, Book VII describes a method, *antanaresis* (now known as the Euclidean algorithm), for finding the greatest common divisor of two or more numbers; Book VIII examines numbers in continued proportions, now known as geometric sequences (such as ax, ax^2, ax^3, ax^4 . . .); and Book IX proves that there is an infinite number of primes.

According to Proclus, Books X and XIII incorporate the work of the Pythagorean Theaetetus (*c.* 417–369 BC). Book X comprises roughly a quarter of the *Elements*, which seems disproportionate to the importance of its classification of incommensurable lines and areas (although study of this book would later inspire Johannes Kepler [1571–1630] in his search for a cosmological model).

Books XI–XIII examine three-dimensional figures, in Greek, *stereometria*. Book XI concerns the intersections of planes, lines, and parallelepipeds (solids with parallel parallelograms as opposite faces). Book XII applies Eudoxus' method of exhaustion to prove that the areas of circles are to one another as the squares of their diameters, and that the volumes of spheres are to one another as the cubes of their diameters. Book XIII culminates with the construction of the five Platonic solids in a given sphere.

The unevenness of the several books and the varied mathematical levels may give the impression that Euclid was but an editor of treatises written by other mathematicians. To some extent this is certainly true, although it is probably impossible to ascertain which parts are his own and which were adaptations from his predecessors. Euclid's contemporaries considered his work final and authoritative; if more was to be said, it had to be as commentaries to the *Elements*.

Almost from the time of its writing, the *Elements* exerted a continuous and major influence on human affairs. It was the primary source of geometric reasoning, theorems, and methods, at least until the advent of non-Euclidean geometry in the nineteenth century. It is sometimes said that, other than the Bible, the *Elements* is the most translated, published, and studied of all the books produced in the western world. Euclid may not have been a first-class mathematician, but he set a standard for deductive reasoning and geometric instruction that persisted, practically unchanged, for more than 2,000 years.

ARCHIMEDES (c. 285–c. 212 BC)

The most famous mathematician
and inventor of Ancient Greece.

Archimedes was born in Syracuse, the principal Greek city-state in Sicily. He probably spent some time in Egypt early in his career, but he resided for most of his life in Syracuse and was on intimate terms with its king, Hieron II (reigned c. 270c. 215 BC). Archimedes published his works in the form of correspondence with the principal mathematicians of his time, including the Alexandrian scholars Conon of Samosand Eratosthenes of Cyrene. He played an important role in the defence of Syracuse against the siege laid by the Romans in 213 BC by constructing war machines so effective that they long delayed the capture of the city. When Syracuse eventually fell to the Roman general Marcus Claudius Marcellus in the autumn of 212 or spring of 211 BC, Archimedes was killed in the sack of the city.

Far more details survive about the life of Archimedes than about any other ancient scientist, but they are largely anecdotal, reflecting the impression that his mechanical genius made on the popular imagination. Thus, he is credited with inventing the Archimedes screw, and he is supposed to have made two "spheres" that Marcellus took back to Rome – one a star globe and the other a device (the details of which are uncertain) for mechanically representing the motions of the sun, the moon, and the planets. The story that he determined the proportion of gold and silver in a wreath made for Hieron by weighing it in water is probably true, but the version that has him leaping from the bath in which he supposedly got the idea and running naked through the streets shouting "*Heureka*!" ("I have found it!") is popular embellishment.

According to Plutarch (*c.* AD 46–119), Archimedes had so low an opinion of the kind of practical invention at which he excelled and to which he owed his contemporary fame that he left no written work on such subjects. While it is true that, apart from a dubious reference to a treatise entitled *On Sphere-Making*, all his known works were of a theoretical character, his interest in mechanics nevertheless deeply influenced his mathematical thinking. Not only did he write works on theoretical mechanics and hydrostatics, but his treatise *Method Concerning Mechanical Theorems* shows that he used mechanical reasoning as a heuristic device for the discovery of new mathematical theorems.

Given the magnitude and originality of Archimedes' achievement, the influence of his mathematics in antiquity was rather small. Those of his results that could be simply expressed – such as the formulae for the surface area and volume of a sphere – became mathematical commonplaces, and one of the bounds he established for pi, $^{22}/_7$, was adopted as the usual approximation to it in antiquity and the Middle Ages. Nevertheless, Archimedes' mathematical work was not continued or developed, as far as is known, in any important way in ancient times, despite his hope expressed in *Method* that its publication would enable others to make new discoveries. However, when some of his treatises were translated into Arabic in the late eighth or ninth century, several mathematicians of medieval Islam were inspired to equal or improve on his achievements. This was the case particularly in the determination of the volumes of solids of revolution, but his influence is also evident in the determination of centres of gravity and in geometric construction problems. Thus, several meritorious works by medieval Islamic mathematicians were inspired by their study of Archimedes.

The greatest impact of Archimedes' work on later mathematicians came in the sixteenth and seventeenth centuries with the printing of texts derived from the Greek, and eventually of the Greek text itself, the *Editio Princeps* ("First Edition"), in Basel in 1544. The Latin translation of many of Archimedes' works by Federico Commandino in 1558 contributed greatly to the spread of knowledge of them, which was reflected in the work of the foremost mathematicians and physicists of the time, including Johannes Kepler (1571–1630) and Galileo Galilei (1564–1642). David Rivault's edition and Latin translation (1615) of Archimedes' complete works, including the ancient commentaries, was enormously influential in the work of some of the best mathematicians of the seventeenth century, notably René Descartes (1596–1650) and Pierre de Fermat (1601–65). Without the background of the rediscovered ancient mathematicians, among whom Archimedes was paramount, the development of mathematics in Europe in the period 1550 to 1650 is inconceivable. It is unfortunate that *Method* remained unknown to both Arabic and Renaissance mathematicians (it was only rediscovered in the late nineteenth century), for they might have fulfilled Archimedes' hope that the work would prove useful in the discovery of theorems.

PLINY THE ELDER (AD 23–79)

Roman savant and author
of the celebrated *Natural History*.

Pliny was descended from a prosperous family, and he was able to complete his studies in Rome. At the age of 23 he began a military career by serving in Germany, rising to the rank of cavalry commander. He returned to Rome, where he possibly studied law. Until near the end of Nero's reign, when he became procurator in Spain, Pliny lived in semi-retirement, studying and writing. His devotion to his studies and his research technique were described by his nephew, Pliny the Younger. Upon the accession in AD 69 of Vespasian, with whom Pliny had served in Germany, he returned to Rome and assumed various official positions.

Pliny's last assignment was that of commander of the fleet in the Bay of Naples, where he was charged with the suppression of piracy. Learning of an unusual cloud formation –later found to have resulted from an eruption of Mt Vesuvius – Pliny went ashore to ascertain the cause and reassure the terrified citizens. He was overcome by the fumes resulting from the volcanic activity and died on August 24 AD 79, according to his nephew's report.

Seven writings are ascribed to him, of which only the *Natural History* survives. There endure, however, a few fragments of his earlier writings on grammar, a biography of Pomponius Secundus, a history of Rome, a study of the Roman campaigns in Germany, and a book on hurling the lance. These writings probably were lost in antiquity and have played no role in perpetuating Pliny's fame, which rests solely on the *Natural History*.

The *Natural History*, divided into 37 *libri*, or "books", was completed, except for finishing touches, in AD 77. In the preface, dedicated to Titus (who became emperor shortly before Pliny's death), Pliny justified the title and explained his purpose on utilitarian grounds as the study of "the nature of things, that is, life". Heretofore, he continued, no one had attempted to bring together the older, scattered material that belonged to "encyclic culture" (*enkyklios paideia*, the origin of the word encyclopaedia). Disdaining high literary style and political mythology, Pliny adopted a plain style – but one with an unusually rich vocabulary – as best suited to his purpose. A novel feature of the *Natural History* is the care taken by Pliny in naming his sources, more than 100 of which are mentioned. Book I, in fact, is a summary of the remaining 36 books, listing the authors and sometimes the titles of the books (many of which are now lost) from which Pliny derived his material.

The *Natural History* properly begins with Book II, which is devoted to cosmology and astronomy. Here, as elsewhere, Pliny demonstrated the extent of his reading, especially of Greek texts. By the same token, however, he was sometimes careless in translating details, with the result that he distorted the meaning of many technical and mathematical passages. In Books III to VI, on the physical and historical geography of the ancient world, he gave much attention to major cities, some of which no longer exist.

Books VII to XI treat zoology, beginning with humans (VII), then mammals and reptiles (VIII), fishes and other marine animals (IX), birds (X), and insects (XI). Pliny derived most of the biological data from Aristotle, while his own contributions were concerned with legendary animals and unsupported folklore.

In Books XII to XIX, on botany, Pliny came closest to making a genuine contribution to science, reporting some independent observations – particularly those made during his travels in Germany. Pliny is one of the chief sources of modern knowledge of Roman gardens, early botanical writings, and the introduction into Italy of new horticultural and agricultural species. Book XVIII, on agriculture, is especially important for agricultural techniques such as crop rotation, farm management, and the names of legumes and other crop plants. His description of an ox-driven grain harvester in Gaul, long regarded by scholars as imaginary, was confirmed by the discovery in southern Belgium in 1958 of a stone relief that dates to about the second century AD and depicts such an implement. Moreover, by recording the Latin synonyms of Greek plant names, Pliny made most of the plants mentioned in earlier Greek writings identifiable.

Books XX to XXXII focus on medicine and drugs. Like many Romans, Pliny criticized luxury on moral and medical grounds. His random comments on diet and on the commercial sources and prices of the ingredients of costly drugs provide valuable evidence relevant to contemporary Roman life. The subjects of Books XXXIII to XXXVII include minerals, precious stones, and metals, especially those used by Roman craftsmen. In describing their uses, Pliny referred to famous artists and their creations and to Roman architectural styles and technology.

In retrospect, Pliny's influence is based on his ability to assemble in a methodical fashion a number of previously unrelated facts, his perceptiveness in recognizing details ignored by others, and his readable stories, in which he linked both factual and fictional data. Along with unsupported claims, fables, and exaggerations, Pliny's belief in magic and superstition helped shape scientific and medical theory in subsequent centuries. Perhaps the most important of the pseudoscientific methods advocated by him was the doctrine of signatures: a resemblance between the external appearance of a plant, animal, or mineral and the outward symptoms of a disease was thought to indicate the therapeutic usefulness of the organism or substance. With the decline of the ancient world and the loss of the Greek texts on which Pliny had so heavily depended, the *Natural History* became a substitute for a general education. In the European Middle Ages many of the larger monastic libraries possessed copies of the work; these and many abridged versions ensured Pliny's place in European literature. His authority was unchallenged, partly because of a lack of more reliable information and partly because his assertions were not, and in many cases could not, be tested.

PTOLEMY (*c.* AD 100–*c.* 170)

Egyptian astronomer, mathematician, and geographer, famed for his geocentric (Earth-centred) model of the universe, now known as the Ptolemaic system.

Virtually nothing is known about Ptolemy's life except what can be inferred from his writings. His first major astronomical work, the *Almagest*, was completed in about AD 150 and contains reports of astronomical observations that Ptolemy had made over the preceding quarter of a century.

The *Almagest* was called by Ptolemy *He mathematike syntaxis* ("The Mathematical Collection") because he believed that its subject, the motions of the heavenly bodies, could be explained in mathematical terms. The opening chapters present empirical arguments for the basic cosmological framework within which Ptolemy worked. The Earth, he argued, is a stationary sphere at the centre of a vastly larger celestial sphere that revolves at a perfectly uniform rate around the Earth, carrying with it the stars, planets, sun, and moon – thereby causing their daily risings and settings. Through the course of a year the sun slowly traces out a great circle, known as the ecliptic, against the rotation of the celestial sphere. (The planets similarly travel against the "fixed stars" found in the ecliptic: hence the planets were known as "wandering stars.") The fundamental assumption of the *Almagest* is that the apparently irregular movements of the heavenly bodies are in reality combinations of regular, uniform, circular motions.

How much of the *Almagest* is original is difficult to determine, because almost all of the preceding technical astronomical literature is now lost. Ptolemy credited Hipparchus (*c.* 190–*c.* 127 BC) with essential elements of his solar theory, as well as with parts of his lunar theory, while denying that Hipparchus constructed planetary models. Ptolemy made only a few vague and disparaging remarks regarding theoretical work over the intervening three centuries, yet the study of the planets undoubtedly progressed considerably during that interval. Moreover, Ptolemy's veracity, especially as an observer, has been controversial since the time of the astronomer Tycho Brahe (1546–1601). Brahe pointed out that solar observations Ptolemy claimed to have made in 141 are definitely not genuine, and there are strong arguments for doubting that Ptolemy independently observed the more than 1,000 stars listed in his star catalogue. What is not disputed, however, is the mastery of mathematical analysis that Ptolemy exhibited.

Ptolemy was pre-eminently responsible for the geocentric cosmology that prevailed in the Islamic world and in medieval Europe. This was not due to the *Almagest* so much as a later treatise, *Hypotheseis ton planomenon* ("Planetary Hypotheses"). In this work he proposed what is now called the Ptolemaic system – a unified system in which each heavenly body is attached to its own sphere and the set of spheres nested so that it extends without gaps from the Earth to the celestial sphere. The numerical tables in the *Almagest* (which enabled planetary positions and other celestial phenomena to be calculated for arbitrary dates) had a profound influence on medieval astronomy, in part through a separate, revised version of the tables that Ptolemy published as *Procheiroi kanones* ("Handy Tables"). Ptolemy taught later astronomers how to use dated, quantitative observations to revise cosmo-logical models.

Ptolemy also attempted to place astrology on a sound basis in *Apotelesmatika* ("Astrological Influences"), later known as the *Tetrabiblos* for its four volumes. He believed that astrology is a legitimate, though inexact, science that describes the physical effects of the heavens on terrestrial life. Ptolemy accepted the basic validity of the traditional astrological doctrines, but he revised the details to reconcile the practice with an Aristotelian conception of nature, matter, and change. Of Ptolemy's writings, the *Tetrabiblos* is the most foreign to modern readers, who do not accept astral prognostication and a cosmology driven by the interplay of basic qualities such as hot, cold, wet, and dry.

Ptolemy has a prominent place in the history of mathe-matics, primarily because of the mathematical methods he applied to astronomical problems. His contributions to trigo-nometry are especially important. For instance, Ptolemy's table of the lengths of chords in a circle is the earliest surviving table of a trigonometric function. He also applied fundamental theorems in spherical trigonometry (apparently discovered half a century earlier by Menelaus of Alexandria) to the solution of many basic astronomical problems.

Among Ptolemy's earliest treatises, the *Harmonics* investi-gated musical theory while steering a middle course between an extreme empiricism and the mystical arithmetical specula-tions associated with Pythagoreanism. Ptolemy's discussion of the roles of reason and the senses in acquiring scientific knowledge have bearing beyond music theory.

Probably near the end of his life, Ptolemy turned to the study of visual perception in *Optica* ("Optics"), a work that survives only in a mutilated medieval Latin translation of an Arabic translation. The extent to which Ptolemy subjected visual perception to empirical analysis is remarkable when contrasted with contemporary Greek writers on optics. For example, Hero of Alexandria (*c.* AD 10–*c.* 70) asserted, purely for philosophical reasons, that an object and its mirror image must make equal angles to a mirror. In contrast, Ptolemy established this principle by measuring angles of incidence and reflection for planar and curved mirrors set upon a disk graduated in degrees. Ptolemy also measured how lines of sight are refracted at the boundary between materials of different density, such as air, water, and glass, although he failed to discover the exact law relating the angles of incidence and refraction.

Ptolemy's fame as a geographer is hardly less than his fame as an astronomer. *Geographike hyphegesis* ("Guide to Geography") provided all the information and techniques required to draw maps of the portion of the world known by Ptolemy's contemporaries. By his own admission, Ptolemy did not attempt to collect and sift all the geographical data on which his maps were based. Instead, he based them on the maps and writings of Marinus of Tyre (*c.* AD 70–*c.* 130), only selectively introducing more current information, chiefly concerning the Asian and African coasts of the Indian Ocean. Nothing would be known about Marinus if Ptolemy had not preserved the substance of his cartographical work.

Ptolemy's most important geographical innovation was to record longitudes and latitudes in degrees for roughly 8,000 locations on his world map, making it possible to make an exact duplicate of the map. Hence, we possess a clear and detailed image of the inhabited world as it was known to a resident of the Roman Empire at its height – a world that extended from the Shetland Islands in the north to the sources of the Nile in the south; from the Canary Islands in the west to China and South-east Asia in the east. Ptolemy's map is seriously distorted in size and orientation compared with modern maps – a reflection of the incomplete and inaccurate descriptions of road systems and trade routes at his disposal.

He also devised two ways of drawing a grid of lines on a flat map to represent the circles of latitude and longitude on the globe. His grid gives a visual impression of the Earth's spherical surface and also, to a limited extent, preserves the proportionality of distances. The more sophisticated of these map projections, using circular arcs to represent both parallels and meridians, anticipated later area-preserving projections. Ptolemy's geographical work was almost unknown in Europe until about 1300, when Byzantine scholars began producing many manuscript copies, several of them illustrated with expert reconstructions of Ptolemy's maps. The Italian Jacopo d'Angelo translated the work into Latin in 1406. The numerous Latin manuscripts and early print editions of Ptolemy's *Guide to Geography*, most of them accompanied by maps, attest to the profound impression this work made upon its rediscovery by Renaissance humanists.

GALEN OF PERGAMUM (AD 129–c. 216)

Greek physician, writer, and philosopher who influenced medical theory and practice in Europe from the Middle Ages until the mid-seventeenth century.

The son of a wealthy architect, Galen was educated as a philosopher and man of letters. His hometown, Pergamum, Anatolia (now Bergama, Turkey), was the site of a magnificent shrine of the healing god, Asclepius, that was visited by many distinguished figures of the Roman Empire for cures. When Galen was 16, he changed his career to that of medicine, which he studied at Pergamum, at Smyrna (modern Izmir, Turkey), and finally at Alexandria in Egypt, which was the greatest medical centre of the ancient world. After more than a decade of study, he returned in AD 157 to Pergamum, where he served as chief physician to the troop of gladiators maintained by the high priest of Asia.

In 162 the ambitious Galen moved to Rome. There he quickly rose in the medical profession owing to his public demonstrations of anatomy, his successes with rich and influential patients whom other doctors had pronounced incurable, his enormous learning, and the rhetorical skills he displayed in public debates. Galen's wealthy background, social contacts, and a friendship with his old philosophy teacher Eudemus, further enhanced his reputation as a philosopher and physician.

Galen abruptly ended his sojourn in the capital in 166. Although he claimed that the intolerable envy of his colleagues prompted his return to Pergamum, an impending plague in Rome was probably a more compelling reason. In 168–9, however, he was called by the joint emperors Lucius Verus and Marcus Aurelius to accompany them on a military campaign in northern Italy. After Verus's sudden death in 169 Galen returned to Rome, where he served Marcus Aurelius and the later emperors Commodus and Septimius Severus as a physician. Galen's final works were written after 207.

Galen regarded anatomy as the foundation of medical knowledge, and he frequently dissected and experimented on such lower animals as the Barbary ape (or African monkey), pigs, sheep, and goats. Galen's advocacy of dissection, both to improve surgical skills and for research purposes, formed part of his self-promotion, but there is no doubt that he was an accurate observer. He distinguished seven pairs of cranial nerves, described the valves of the heart, and observed the structural differences between arteries and veins. One of his most important demonstrations was that the arteries carry blood, not air, as had been taught for 400 years.

Galen's physiology was a mixture of ideas taken from the philosophers Plato and Aristotle as well as from the physician Hippocrates, whom Galen revered as the font of all medical learning. Galen viewed the body as consisting of three connected systems: the brain and nerves, which are responsible for sensation and thought; the heart and arteries, responsible for life-giving energy; and the liver and veins, responsible for nutrition and growth. According to Galen, blood is formed in the liver and is then carried by the veins to all parts of the body, where it is used up as nutriment or is transformed into flesh and other substances.

Building on earlier Hippocratic conceptions, Galen believed that human health requires an equilibrium between the four main bodily fluids, or humours – blood, yellow bile, black bile, and phlegm. Each of the humours is built up from the four elements and displays two of the four primary qualities: hot, cold, wet, and dry. Unlike Hippocrates, Galen argued that humoral imbalances can be located in specific organs, as well as in the body as a whole. This modification of the theory allowed doctors to make more precise diagnoses and to prescribe specific remedies to restore the body's balance. As a continuation of earlier Hippocratic conceptions, Galenic physiology became a powerful influence in medicine for the next 1,400 years.

Galen's writings achieved wide circulation during his lifetime, and copies of some of his works survive that were written within a generation of his death. By AD 500 his works were being taught and summarized at Alexandria, and his theories were already crowding out those of others in the medical handbooks of the Byzantine world. Greek manuscripts began to be collected and translated by enlightened Arabs in the ninth century, and in about 850 Hunayn ibn Ishaq, an Arab physician at the court of Baghdad, prepared an annotated list of 129 works of Galen that he and his followers had translated from Greek into Arabic or Syriac.

However, in western Europe Galen's influence was almost negligible until the late eleventh century, when his work began to be translated into Latin. These Latin versions came to form the basis of medical education in the new medieval universities. From about 1490, Italian humanists felt the need to prepare new Latin versions of Galen directly from Greek manuscripts in order to free his texts from medieval preconceptions and misunderstandings. Galen's works were first printed in Greek in their entirety in 1525, and printings in Latin swiftly followed. These texts offered a different picture from those of the Middle Ages, one that emphasized Galen as a clinician, a diagnostician, and above all an anatomist, and his injunctions to investigate the body were eagerly followed. Paradoxically, this soon led to the overthrow of Galen's authority as an anatomist. In 1543 the Flemish physician Andreas Vesalius showed that Galen's anatomy of the body was more animal than human in some of its aspects, and it became clear that Galen and his medieval followers had made many errors.

AL-KHWARIZMI (*c.* 780–*c.* 850)

Muslim mathematician and astronomer whose major works introduced Hindu-Arabic numerals and the concepts of algebra into European mathematics.

Al-Khwārizmī was born and lived in Baghdad, where he worked at the *Dār al-Ḥīkma* ("House of Wisdom") under the caliphate of al-Ma'mūn. The House of Wisdom acquired and translated scientific and philosophical treatises, particularly Greek, as well as publishing original research. Al-Kwārizmī's work on elementary algebra, *al-Kitāb al-mukhtaṣar fī ḥisāb al-jabr wa'l-muqābala* ("The Compendious Book on Calculation by Completion and Balancing"), was translated into Latin in the twelfth century, from which the title and term *Algebra* derives. *Algebra* is a compilation of rules, together with demonstrations, for finding solutions of linear and quadratic equations based on intuitive geometric arguments, rather than the abstract notation now associated with the subject. Its systematic, demonstrative approach distinguishes it from earlier treatments of the subject. It also contains sections on calculating areas and volumes of geometric figures and on the use of algebra to solve inheritance problems according to proportions prescribed by Islamic law. Elements within the work can be traced from Babylonian mathematics of the early second millennium BC through Hellenistic, Hebrew, and Hindu treatises.

In the twelfth century a second work by al-Kwārizmī introduced Hindu-Arabic numerals and their arithmetic to the West. It is preserved only in a Latin translation, *Algoritmi de numero Indorum* ("Al-Khwārizmī Concerning the Hindu Art of Reckoning"). From the name of the author, rendered in Latin as *Algoritmi*, originated the term *algorithm*.

A third major book was his *Kitāb ṣūrat al-Arḍ* ("The Book Image of the Earth"; translated as *Geography*), which presented the coordinates of localities in the known world based, ultimately, on those in the *Geography* of Ptolemy (*c.* AD 100–*c.* 170) but with improved values for the length of the Mediterranean Sea and the location of cities in Asia and Africa. He also assisted in the construction of a world map for al-Ma'mūn and participated in a project to determine the circumference of the Earth, which had long been known to be spherical, by measuring the length of a degree of a meridian through the plain of Sinjār in Iraq.

Finally, al-Khwārizmī also compiled a set of astronomical tables (Zīj), based on a variety of Hindu and Greek sources. This work included a table of sines, evidently for a circle of radius 150 units. Like his treatises on algebra and Hindu-Arabic numerals, this astronomical work (or an Andalusian revision thereof) was translated into Latin.

AVICENNA (980–1037)

The most famous and influential
of the philosopher-scientists of Islam.

Avicenna, a Persian who spent his whole life in the eastern and central regions of Iran, was born in Bukhara, where he received his earliest education under the direction of his father. Since the house of his father was a meeting place for learned men, from early childhood Avicenna was able to profit from the company of the outstanding masters of his day. A precocious child with an exceptional memory, which he retained throughout his life, he had memorized the Qur'ān and much Arabic poetry by the age of ten. He then studied logic and metaphysics under teachers whom he soon outpaced, and then spent the few years until he reached the age of 18 in self-education. He read avidly and mastered Islamic law, then medicine, and finally metaphysics. Particularly helpful in his intellectual development was his gaining access to the rich royal library of the Sāmānids – the first great native dynasty that arose in Iran after the Arab conquest – as the result of his successful cure of the Sāmānid prince Nūḥ ibn Manṣūr. By the time Avicenna was 21 he was accomplished in all branches of formal learning and had already gained a wide reputation as an outstanding physician. His services were also sought as an administrator, and for a while he even entered government service as a clerk.

But suddenly the whole pattern of his life changed. His father died; the Sāmānid house was defeated by Maḥmūd of Ghazna, the Turkish leader and legendary hero who established Ghaznavid rule in Khorāsān (north-eastern Iran and modern western Afghanistan); and Avicenna began a period of wandering and turmoil that, with the exception of a few intervals of tranquillity, was to last to the end of his life. But his powers of concentration and intellectual prowess were such that he was able to continue his intellectual work with remarkable consistency and continuity, and was not at all influenced by the outward disturbances.

Avicenna wandered for a while in different cities of Khorāsān and then left for the court of the Būyid princes, who were ruling over central Iran – first going to Rayy (near modern Tehran) and then to Qazvīn, where as usual he made his living as a physician. But in these cities also he found neither sufficient social and economic support nor the necessary peace and calm to continue his work. He went, therefore, to Hamadan in west-central Iran, where Shams al-Dawlah, another Būyid prince, was ruling. This journey marked the beginning of a new phase in Avicenna's life. He became court physician and enjoyed the favour of the ruler to the extent that twice he was appointed vizier. As was the order of the day, he also suffered political reactions and intrigues against him and was forced into hiding for some time; at one time he was even imprisoned.

In Hamadan Avicenna began his two most famous works. *Kitāb al-shifā'* ("Book of Healing") is probably the largest work of its kind ever written by one man. It treats of logic; the natural sciences, including psychology; the *quadrivium* (geometry, astronomy, arithmetic, and music); and metaphysics. Avicenna's thought in this work owes a great deal to Aristotle but also to other Greek influences. His system rests on the conception of God as the necessary existent: in God alone essence (*what* he is) and existence (*that* he is) coincide. *Al-Qānūn fī al-ṭibb* ("The Canon of Medicine") is the most famous single book in the history of medicine in both the East and the West. It is a systematic encyclopaedia based for the most part on the achievements of Greek physicians of the Roman imperial age and on other Arabic works, and, to a lesser extent, on Avicenna's own experience (his clinical notes were lost during his journeys). Occupied during the day with his duties at court as both physician and administrator, Avicenna spent almost every night with his students composing these and other works and carrying out general philosophical and scientific discussions related to them. These sessions were often combined with musical performances and gaiety and lasted until late in the night. Even in hiding and in prison he continued to write. The great physical strength of Avicenna enabled him to carry out a programme that would have been unimaginable for a person of a weaker constitution.

The last phase of Avicenna's life began with his move to Eṣfahān (about 250 miles south of Tehran). In 1022 Shams al-Dawlah died, and Avicenna, after a period of difficulty that included imprisonment, fled to Eṣfahān with a small entourage. There he was to spend the last 14 years of his life in relative peace. He was esteemed highly by 'Alā' al-Dawlah, the ruler, and his court. Here Avicenna finished the two major works he began in Hamadan and wrote most of his treatises, which amounted to nearly 200. Accompanying 'Alā' al-Dawlah on a campaign, Avicenna fell ill and, despite his attempts to treat himself, died from colic and exhaustion.

Avicenna's influence was felt in the western world, though no distinct school of "Latin Avicennism" can be discerned. Avicenna's *Book of Healing* was translated partially into Latin in the twelfth century, and the complete *Canon* appeared in the same century. These translations and others spread the ideas of Avicenna far and wide in the West. His thought, blended with that of St Augustine, the Christian philosopher and theologian, was a basic ingredient in the doctrines of many of the medieval Scholastics, especially in the Franciscan schools. In medicine the *Canon* became the medical authority for several centuries, and Avicenna enjoyed an undisputed place of honour equalled only by the early Greek physicians Hippocrates and Galen. In the East his dominating influence in medicine, philosophy, and theology has lasted over the ages and is still alive within the circles of Islamic thought.

ROGER BACON (*c.* 1220–1292)

English philosopher and educational reformer
who was a major medieval proponent of experimental
science and become known as Doctor Mirabilis
(Latin for "Wonderful Teacher").

Bacon's birthplace is believed to have been Ilchester, Somerset, or Bisley, Gloucester. Born into a wealthy family, he was well versed in the classics and enjoyed the advantages of an early training in geometry, arithmetic, music, and astronomy. At the beginning of his career, Bacon was a lecturer in the Faculty of Arts in Paris. About 1247, however, a considerable change took place in his intellectual development and from that time he spent much time and energy, and huge sums of money, in experimental research, in acquiring "secret" books, in the construction of instruments and tables, in the training of assistants, and in seeking the friendship of savants. The change was probably caused by his move to Oxford and the influence there of the great scholar Robert Grosseteste, a leader in introducing Greek learning to the West, and his student Adam de Marisco, as well as that of Thomas Wallensis, the bishop of St David's. From 1247 to 1257 Bacon devoted himself whole-heartedly to the cultivation of those new branches of learning to which he was introduced at Oxford – languages, optics, and alchemy – and to further studies in astronomy and mathe-matics. Here he developed an insistence on fruitful lines of research and methods of experimental study.

Bacon's studies on the nature of light and on the rainbow are especially noteworthy, and he seems to have planned and interpreted these experiments carefully. But his most notable "experiments" seem never to have been actually performed; they were merely described. He was the first person in the West to give exact directions for making gunpowder, in 1242, and although he knew that, if confined, it would have great power and might be useful in war, he failed to speculate further. (Its use in guns arose early in the following century.) Bacon elucidated the principles of reflection, refraction, and spherical aberration, and proposed mechanically propelled ships and carriages. He also used a camera obscura (which projects an image through a pinhole) to observe eclipses of the sun. By 1257 Bacon had entered the Order of Friars Minor. Because of ill health and his life in the order, Bacon felt (as he wrote) forgotten by everyone and all but buried. His university and literary careers seemed finished. His feverish activity, his amazing credulity, his superstition, and his vocal contempt for those not sharing his interests displeased his superiors in the order and brought him under severe discipline. He decided to appeal to Pope Clement IV, whom he may have known when the latter was (before his election to the papacy) in the service of the Capetian kings of France.

In a letter in 1266, the pope referred to letters received from Bacon, who had come forward with certain proposals covering the natural world, mathematics, languages, perspective, and astrology. Bacon had argued that a more accurate experimental knowledge of nature would be of great value in confirming the Christian faith, and he felt that his proposals would be of considerable importance for the welfare of the church and of the universities. The pope desired to be more fully informed of these projects and commanded Bacon to send him the work. But Bacon had had in mind a vast encyclopaedia of all the known sciences, requiring many collaborators, the organization and administration of which would be coordinated by a papal institute. The work, then, was merely projected when the pope thought that it already existed.

In obedience to the pope's command, however, Bacon set to work and in a remarkably short time had dispatched the *Opus majus* ("Great Work"), the *Opus minus* ("Lesser Work"), and the *Opus tertium* ("Third Work"). He had to do this secretly and notwithstanding any command of his superiors to the contrary.

Under the circumstances, his achievement was truly astounding. He aspired to penetrate realms undreamed of in the schools at Paris and to lay bare the secrets of nature by positive study. The *Opus majus* was an effort to persuade the pope of the urgent necessity and manifold utility of the reforms that he proposed. But in 1268 Clement died, thereby extinguishing Bacon's dreams of gaining for the sciences their rightful place in the curriculum of university studies. In 1272 he produced the *Compendium philosophiae* ("Compendium of Philosophy"). Bacon's philosophy – and even his so-called scientific works contain lengthy philosophical digressions – was essentially Aristotelian, although it did include Neoplatonist elements.

Sometime between 1277 and 1279, Bacon was condemned to prison by his fellow Franciscans because of certain "suspected novelties" in his teaching. The condemnation was probably issued because of his bitter attacks on the theologians and scholars of his day, his excessive credulity in alchemy and astrology, and his penchant for millenarianism under the influence of the prophecies of Abbot Joachim of Fiore, a mystical philosopher of history. How long he was imprisoned is unknown. His last work, *Compendium studii theologiae* ("Compendium of the Study of Theology"; 1292), incomplete as so many others, shows him as aggressive as ever.

LEONARDO DA VINCI (1452–1519)

Italian painter, draftsman, sculptor, architect, and engineer whose genius, perhaps more than that of any other figure, epitomized the Renaissance humanist ideal.

Leonardo was born in Anchiano, near Vinci, Republic of Florence (now Italy). He received the usual elementary education of that day: reading, writing, and arithmetic. Leonardo did not seriously study Latin, the key language of traditional learning, until much later, when he acquired a working knowledge of it on his own. Leonardo's artistic inclinations must have appeared early. When he was about 15, his father, who enjoyed a high reputation in the Florence community, apprenticed him to artist Andrea del Verrocchio. In Verrocchio's renowned workshop Leonardo received an all-round training that included painting and sculpture as well as the technical-mechanical arts. In 1472 Leonardo was accepted into the painters' guild of Florence, but he remained in his teacher's workshop for five more years, after which time he worked independently in Florence until 1481.

In 1482 Leonardo moved to Milan to work in the service of the city's duke – a surprising step when one considers that the 30-year-old artist had just received his first substantial commissions from his native city of Florence: the unfinished panel painting *The Adoration of the Magi* for the monastery of San Donato a Scopeto and an altar painting for the St Bernard Chapel in the Palazzo della Signoria, which was never begun. He was no doubt enticed by Duke Ludovico Sforza's brilliant court and the meaningful projects awaiting him there.

Leonardo spent 17 years in Milan, until Ludovico's fall from power in 1499. He was listed in the register of the royal household as *pictor et ingeniarius ducalis* ("painter and engineer of the duke"). Leonardo's gracious but reserved personality and elegant bearing were well received in court circles. Highly esteemed, he was kept constantly busy as a painter and sculptor and as a designer of court festivals. He was also frequently consulted as a technical adviser in the fields of architecture, fortifications, and military matters, and he served as a hydraulic and mechanical engineer.

As a painter, Leonardo completed six works in the 17 years in Milan, including the altar painting *The Virgin of the Rocks* (*c.* 1483–6), and the monumental wall painting *The Last Supper* (1495–8) in the refectory of the monastery of Santa Maria delle Grazie. Also of note is the decorative ceiling painting (1498) he made for the Sala delle Asse in the Milan Castello Sforzesco.

The great programme of Leonardo the writer, author of treatises, began between 1490 and 1495. During this period, his interest in two fields – the artistic and the scientific – developed and shaped his future work, building toward a kind of creative dualism that sparked his inventiveness in both fields. He gradually gave shape to four main themes that were to occupy him for the rest of his life: a treatise on painting, a treatise on architecture, a book on the elements of mechanics, and a broadly outlined work on human anatomy. His geophysical, botanical, hydrological, and aerological researches also started in this period. He scorned speculative book knowledge, favouring instead the irrefutable facts gained from experience. An artist by disposition and endowment, he considered his eyes to be his main avenue to knowledge: to Leonardo, sight was man's highest sense because it alone conveyed the facts of experience immediately, correctly, and with certainty. Hence, every phenomenon perceived became an object of knowledge, and *saper vedere* ("knowing how to see") became the great theme of his studies.

It was during his first years in Milan that Leonardo began the earliest of his notebooks. He would first make quick sketches of his observations on loose sheets or on tiny paper pads he kept in his belt, then would arrange them according to theme and enter them in order in the notebook. Surviving in notebooks from throughout his career are a first collection of material for a painting treatise, a model book of sketches for sacred and profane architecture, a treatise on elementary theory of mechanics, and the first sections of a treatise on the human body. Leonardo's notebooks add up to thousands of closely written pages abundantly illustrated with sketches – the most voluminous literary legacy any painter has ever left behind. Of more than 40 codices mentioned – sometimes inaccurately – in contemporary sources, 21 have survived; these in turn sometimes contain notebooks originally separate but now bound, so that 32 in all have been preserved.

One special feature that makes Leonardo's notes and sketches unusual is his use of mirror writing. He was left-handed, so mirror writing came easily and naturally to him – although it is unclear why he chose to use this. While somewhat idiosyncratic, his script can be read clearly and without difficulty with the help of a mirror – as his contemporaries testified – and should not be looked on as a secret handwriting. But the fact that Leonardo used mirror writing throughout his notebooks, even in the copies drawn up with painstaking calligraphy, indicates that – although he constantly addressed an imaginary reader in his writings – he never felt the need to achieve easy communication. His writings must be interpreted as preliminary stages of works destined for eventual publication that he never got round to completing. In a sentence in the margin of one of his late anatomy sketches, Leonardo implores his followers to see that his works are printed.

In December 1499 or, at the latest, January 1500 – shortly after the victorious entry of the French into Milan – Leonardo left that city in the company of the mathematician Lucas Pacioli. He returned to Florence, where, after a long absence, he was received with acclaim and honoured as a renowned native son. In the same year he was appointed an architectural expert on a committee investigating damages to the foundation and structure of the church of San Francesco al Monte. Leonardo, however, seems to have been concentrating more on mathematical studies than painting. He left Florence in the summer of 1502 to enter the service of Cesare Borgia as "senior military architect and general engineer". Borgia, the notorious son of Pope Alexander VI, had, as commander in chief of the papal army, sought with unparalleled ruthlessness to gain control of the Papal States of Romagna and the Marches. In the spring of 1503 Leonardo returned to Florence. He received a prized commission to paint the *Battle of*

Anghiari, a large mural for the council hall in Florence's Palazzo Vecchio. During this time he also painted the *Mona Lisa,* one of the most popular and most analysed paintings of all time. The sense of harmony achieved in the painting – especially apparent in the sitter's faint smile – reflects Leonardo's idea of the cosmic link connecting humanity and nature.

In Florence Leonardo did dissections in the hospital of Santa Maria Nuova and broadened his anatomical work into a comprehensive study of the structure and function of the human organism. He made systematic observations of the flight of birds, about which he planned a treatise. Even his hydrological studies broadened into research on the physical properties of water, especially the laws of currents, which he compared to those pertaining to air. These were also set down in his own collection of data, contained in the so-called "Codex Hammer" (formerly known as the Leicester Codex).

In May 1506 Charles d'Amboise, the French governor in Milan, asked the Signoria in Florence if Leonardo could travel to Milan. The Signoria let Leonardo go, and the monumental *Battle of Anghiari* remained unfinished. Honoured and ad-mired by his generous patrons in Milan, Charles d'Amboise and King Louis XII, Leonardo enjoyed his duties, which were limited largely to advice in architectural matters. Leonardo's scientific activity flourished during this period. His studies in anatomy achieved a new dimension in his collaboration with Marcantonio della Torre, a famous anatomist from Pavia. Leonardo outlined a plan for an overall work that would include not only exact, detailed reproductions of the human body and its organs but also comparative anatomy and the whole field of physiology. Leonardo's manuscripts of the period are replete with mathematical, optical, mechanical, geological, and botanical studies. These investigations became increasingly driven by a central idea: the conviction that force and motion as basic mechanical functions produce all outward forms in organic and inorganic nature and give them their shape. Furthermore, he believed that these functioning forces operate in accordance with orderly, harmonious laws.

In 1513 political events – the temporary expulsion of the French from Milan – caused the now 60-year-old Leonardo to move again. At the end of the year he went to Rome. However, perhaps stifled by this scene, at age 65 Leonardo accepted the invitation of the young King Francis I to enter his service in France. At the end of 1516 he left Italy forever. Leonardo spent the last three years of his life in the small residence of Cloux (later called Clos-Lucé), near the King's summer palace at Amboise in the Loire valley. Leonardo did little painting while in France, spending most of his time arranging and editing his scientific studies, his treatise on painting, and a few pages of his anatomy treatise. In the so-called *Visions of the End of the World*, or *Deluge*, series (*c.* 1514–15), he depicted with an overpowering imagination the primal forces that rule nature, while also perhaps betraying his growing pessimism. Leonardo died at Cloux and was buried in the palace church of Saint-Florentin.

As the fifteenth century expired, Scholastic doctrines were in decline and humanistic scholarship was on the rise. Leonardo, however, was part of an intellectual circle that developed a third, specifically modern, form of cognition. In his view, the artist – as transmitter of the true and accurate data of experience acquired by visual observation – played a significant part. In an era that often compared the process of divine creation to the activity of an artist, Leonardo reversed the analogy, using art as his own means to approximate the mysteries of creation: asserting that, through the science of painting, "the mind of the painter is transformed into a copy of the divine mind, since it operates freely in creating many kinds of animals, plants, fruits, landscapes, countrysides, ruins, and awe-inspiring places." With this sense of the artist's high calling, Leonardo approached the vast realm of nature to probe its secrets. His utopian idea of transmitting in encyclopaedic form the knowledge thus gained was still bound up with medieval Scholastic conceptions, but the results of his research were among the first great achievements of the forthcoming age, because they were based to an unprecedented degree on the principle of experience.

Although he made strenuous efforts to become erudite in languages, natural science, mathematics, philosophy, and history, as a mere listing of the wide-ranging contents of his library demonstrates, Leonardo remained an empiricist of visual observation. It is precisely through this observation – and his own genius – that he developed a unique "theory of knowledge" in which art and science form a synthesis. In the face of his overall achievements, therefore, the question of how much he finished or did not finish is irrelevant. The crux of the matter is his intellectual force, inherent in every one of his creations – a force that continues to spark scholarly interest today.

NICOLAUS COPERNICUS (1473–1543)

Polish astronomer who proposed the heliocentric ("sun-centred") model of the heavens.

Copernicus was born in Torun, a city in north-central Poland on the Vistula River south of the major Baltic seaport of Gdansk. His father, Nicolaus, was a well-to-do merchant, and his mother, Barbara Watzenrode, also came from a leading merchant family. Between 1491 and about 1494 Copernicus studied liberal arts – including astronomy and astrology – at the University of Cracow (Kraków). He left before completing his degree and resumed his studies in Italy at the University of Bologna. For a time he lived in the same house as the principal astronomer at the university, Domenico Maria de Novara (Latin: Domenicus Maria Novaria Ferrariensis; 1454–1504), responsible for issuing annual astrological prognostications for the city. Copernicus acted as "assistant and witness" to some of Novara's observations, and his involvement with the production of the annual forecasts means that he was intimately familiar with the practice of astrology. Novara also probably introduced Copernicus to early works that criticized the Ptolemaic, or geocentric ("Earth-centred"), model of the heavens.

In 1500 Copernicus spoke before an interested audience in Rome on mathematical subjects, but the exact content of his lectures is unknown. In 1501 he stayed briefly in Frauenburg but soon returned to Italy to continue his education, this time at the University of Padua, where he pursued medical studies until 1503. In May of that year he finally received a doctorate in canon law, but from an Italian university where he had not studied: the University of Ferrara.

When Copernicus returned to Poland, Bishop Watzenrode arranged a sinecure for him: an *in absentia* teaching post, or scholastry, at Wrocław. Copernicus's actual duties at the bishopric palace, however, were largely administrative and medical. As a church canon, he collected rents from church-owned lands; secured military defences; oversaw chapter finances; managed the bakery, brewery, and mills; and cared for the medical needs of the other canons and his uncle. His astronomical work took place in his spare time. It was during the last years of Watzenrode's life that Copernicus came up with the idea on which his subsequent fame was to rest – that of a heliocentric ("sun-centred") model of the heavens. His theory had important consequences for later thinkers of the scientific revolution, including such major figures as Galileo (1564–1642), Kepler (1571–1630), Descartes (1596–1650), and Newton (1642–1727).

From antiquity, there was general agreement that the moon and sun encircled the motionless Earth and that Mars, Jupiter, and Saturn were situated beyond the sun in that order. Mercury and Venus were also believed to encircle the Earth, although there was disagreement concerning their location relative to the sun. Astronomical modelling was governed by the premise that these bodies move with uniform angular motion on fixed radii at a constant distance from their centres of motion.

In the *Commentariolus* ("Little Commentary"), written between 1508 and 1514, Copernicus postulated that, if the sun is assumed to be at rest and if the Earth is assumed to be in motion, then the planets fall into an orderly relationship whereby their sidereal periods (the time taken to complete an orbit) increase from the sun as follows: Mercury (88 days), Venus (225 days), Earth (1 year), Mars (1.9 years), Jupiter (12 years), and Saturn (30 years). This theory did resolve the disagreement about the ordering of the planets, but it raised new problems.

To accept the theory's premises, one had to abandon much of Aristotelian natural philosophy and develop a new explanation for why heavy bodies fall to a moving Earth. It was also necessary to explain how a transient body like the Earth, filled with meteorological phenomena, pestilence, and wars, could be part of a perfect and imperishable heaven. In addition, Copernicus was working with many observations that he had inherited from antiquity and whose trustworthiness he could not verify. In constructing a theory for the precession of the equinoxes (a gradual change in the orientation of the Earth's axis), for example, he was trying to build a model based upon very small, long-term effects. And his theory for Mercury was left with serious incoherencies.

Any of these considerations alone could account for Copernicus's delay in publishing until 1543 the final version of his theory in *De revolutionibus orbium coelestium libri vi* ("Six Books Concerning the Revolutions of the Heavenly Orbs"). A description of the main elements of the heliocentric hypothesis was first published, in the *Narratio prima* ("First Narration"; 1540–1), not under Copernicus's name but under that of the 25-year-old Georg Rheticus. The *Narratio prima* was, in effect, a joint production of Copernicus and Rheticus, something of a "trial balloon" for the main work. It provided a summary of the theoretical principles contained in the manuscript of *De revolutionibus*, emphasized their value for computing new planetary tables, and presented Copernicus as following admiringly in the footsteps of Ptolemy even as he broke fundamentally with his ancient predecessor. It also provided what was missing from the *Commentariolus:* a basis for accepting the claims of the new theory. Both Rheticus and Copernicus knew that they could not definitively rule out all possible alternatives to the heliocentric theory. But they could underline what Copernicus's theory provided that others could not: a singular method for ordering the planets and for calculating the relative distances of the planets from the sun. Rheticus compared this new universe to a well-tuned musical instrument and to the interlocking wheel mechanisms of a clock.

The presentation of Copernicus's theory in its final form is inseparable from the conflicted history of its publication. When Rheticus left Frauenburg to return to his teaching duties at Wittenberg, he took the manuscript with him in order to arrange for its publication at Nürnberg, the leading centre of printing in Germany. He chose the top printer in the city, Johann Petreius. It was not uncommon for authors to participate directly in the printing of their manuscripts, sometimes even living in the printer's home. However, Rheticus was unable to remain and supervise. He turned the manuscript over to Andreas Osiander (1498–1552), a theologian experienced in shepherding mathematical books through production as well as a leading political figure in the city and an ardent Lutheran. In earlier communication with Copernicus, Osiander had urged him to present his ideas as purely hypothetical, and he now introduced certain changes without the permission of either Rheticus or Copernicus. Osiander added an unsigned "Letter to the Reader" directly after the title page, which maintained that the hypotheses contained within made no pretence to truth and that, in any case, astronomy was incapable of finding the causes of heavenly phenomena. A casual reader would be confused about the relationship between this letter and the book's contents. Both Petreius and Rheticus, having trusted Osiander, now found themselves betrayed.

Legend has it that a copy of *De revolutionibus* was placed in Copernicus's hands a few days after he lost consciousness from a stroke. He awoke long enough to realize that he was holding his great book and then expired, "publishing as he perished". In fact, he died on 24 May 1543, some two months after publication. His legend overlooks this time lapse, giving it the beatific air of a saint's tale.

PARACELSUS (1493–1541)

German-Swiss physician and alchemist who
established the role of chemistry in medicine.

Paracelsus was born in Einsiedeln, Switzerland, the only son of a somewhat impoverished German doctor and chemist.

Theophrastus, as he was first called, was a small boy when his mother died; his father then moved to Villach in southern Austria. In 1507, at the age of 14, Paracelsus joined the many vagrant youths who swarmed across Europe in the late Middle Ages, seeking famous teachers at one university after another. During the next five years he is said to have attended the universities of Basel, Tübingen, Vienna, Wittenberg, Leipzig, Heidelberg, and Cologne, but was disappointed with them all. He later wrote that he wondered how "the high colleges managed to produce so many high asses", a typical Paracelsian jibe.

His attitude upset the schoolmen. "The universities do not teach all things," he wrote, "so a doctor must seek out old wives, gipsies, sorcerers, wandering tribes, old robbers, and such outlaws and take lessons from them. A doctor must be a traveller, . . . Knowledge is experience." Paracelsus held that the rough-and-ready language of the innkeeper, barber, and teamster had more real dignity and common sense than the dry-as-dust scholasticism of Aristotle, Galen, and Avicenna, the recognized Greek and Arab medical authorities of his day.

Paracelsus is said to have graduated from the University of Vienna with the baccalaureate in medicine in 1510, when he was 17. He was, however, delighted to find the medicine of Galen and the medieval Arab teachers criticized in the University of Ferrara, where, he always insisted, he received his doctoral degree in 1516 (university records are missing for that year). At Ferrara he was free to express his rejection of the prevailing view that the stars and planets controlled all the parts of the human body. He is thought to have begun using the name "para-Celsus" (i.e. above or beyond Celsus) at about that time, for he regarded himself as even greater than Celsus, the renowned first-century Roman physician.

Soon after taking his degree, he set out upon many years of wandering through almost every country in Europe, including

England, Ireland, and Scotland. He then took part in the "Netherlandish wars" as an army surgeon, at that time a lowly occupation. Later he went to Russia, was held captive by the Tatars, escaped into Lithuania, went south into Hungary, and in 1521 again served as an army surgeon, in Italy. Ultimately his wanderings brought him to Egypt, Arabia, the Holy Land, and, finally, Constantinople. Everywhere he sought out the most learned exponents of practical alchemy, not only to discover the most effective means of medical treatment but also – and more importantly – to discover "the latent forces of Nature".

After about ten years of wandering, he returned home to Villach in 1524 to find that his fame for many miraculous cures had preceded him. When it became known that the Great Paracelsus, then aged 33, had been appointed town physician and lecturer in medicine at the University of Basel, students from all parts of Europe began to flock into the city. Pinning a programme of his forthcoming lectures to the noticeboard of the university on 5 June 1527, he invited not only students but everyone and anyone. The authorities were scandalized and incensed by his open invitation.

Three weeks later, on 24 June 1527, surrounded by a crowd of cheering students, he burned the books of Avicenna, the Arab "Prince of Physicians", and those of the Greek physician Galen, in front of the university. No doubt his enemies recalled how Luther, just six and a half years before on December 10 1520 at the Elster Gate of Wittenberg, had burned a papal bull. Like Luther, Paracelsus also lectured and wrote in German rather than Latin, for he loved the common tongue.

Despite his bombastic blunders, he reached the peak of his tempestuous career at Basel. His name and fame spread throughout the known world, and his lecture hall was crowded to overflowing. He stressed the healing power of nature and raged against those methods of treating wounds, such as padding with moss or dried dung, that prevented natural draining. The wounds must drain, he insisted, for "If you prevent infection, Nature will heal the wound all by herself." He attacked venomously many other medical malpractices of his time and jeered mercilessly at worthless pills, salves, infusions, balsams, electuaries, fumigants, and drenches, much to the delight of his student-disciples.

Paracelsus's triumph at Basel lasted less than a year, however, for he had made too many enemies. By the spring of 1528 he was at loggerheads with doctors, apothecaries, and magistrates. Finally, and suddenly, he had to flee for his life in the dead of night. Alone and penniless, he wandered toward Colmar in Upper Alsace. He stayed at various places with friends, and such leisurely travel for the next eight years allowed him to revise old manuscripts and write new treatises. With the publication of *Der grossen Wundartzney* ("Great Surgery Book") in 1536 he made an astounding comeback: this book restored, and even extended, the almost fabulous reputation he had earned at Basel in his heyday.

The medical achievements of Paracelsus were outstanding. In 1530 he wrote the best clinical description of syphilis up to that time, maintaining that it could be successfully treated by carefully measured doses of mercury compounds taken internally. He stated that the "miners' disease" (silicosis) resulted from inhaling metal vapours and was not a punishment for sin administered by mountain spirits. He was the first to declare that, if given in small doses, "what makes a man ill also cures him", an anticipation of the modern practice of homeopathy. Paracelsus is said to have cured many persons in the plague-stricken town of Stertzing in the summer of 1534 by administering orally a pill made of bread containing a minute amount of the patient's excreta he had removed on a needle point.

Paracelsus was the first to connect goitre with minerals, especially lead, in drinking water. He prepared and used new chemical remedies, including those containing mercury, sulphur, iron, and copper sulphate – thus uniting medicine with chemistry, as the first *London Pharmacopoeia*, in 1618, indicates. Carl Jung (1875–1961), the psychiatrist, wrote of him that "We see in Paracelsus not only a pioneer in the domains of chemical medicine, but also in those of an empirical psychological healing science."

ANDREAS VESALIUS (1514–1564)

Renaissance Flemish physician who revolutionized the study of biology and the practice of medicine.

Vesalius was born in Brussels to a family of physicians and pharmacists. He attended the University of Louvain in 1529–33, and from 1533 to 1536 studied at the medical school of the University of Paris, where he learned to dissect animals. He also had the opportunity to dissect human cadavers, and devoted much of his time to a study of human bones – at that time easily available in the Paris cemeteries.

In 1536 he returned to his native Brabant to spend another year at the University of Louvain, where the influence of Arab medicine was still dominant. He then went to the University of Padua, a progressive institution with a strong tradition of anatomical dissection. On receiving the MD degree the same year, he was appointed lecturer in surgery with the responsibility of giving anatomical demonstrations. Since he knew that a thorough knowledge of human anatomy was essential to surgery, he devoted much of his time to dissections of cadavers and insisted on doing them himself, instead of relying on untrained assistants.

At first, Vesalius had no reason to question the theories of Galen, at that time still considered as authoritative in medical education. In January 1540, while visiting the University of Bologna, Vesalius broke with the tradition of relying on Galen and openly demonstrated dissections that he did himself. By learning anatomy from cadavers and critically evaluating ancient texts, Vesalius was soon convinced that the anatomy of Galen had not been based on the dissection of the human body, which had been strictly forbidden by the Roman religion. Galenic anatomy, he maintained, was an application to the human form of conclusions drawn from the dissections of animals: mostly dogs, monkeys, or pigs.

Vesalius soon began to prepare a complete textbook on human anatomy. Early in 1542 he travelled to Venice to supervise the preparation of drawings to illustrate his text, probably in the studio of the great Renaissance artist Titian. The drawings of his dissections were engraved on wood blocks, which he took, together with his manuscript, to Basel, Switzerland, where his major work *De humani corporis fabrica libri septem* ("The Seven Books on the Structure of the Human Body") commonly known as the *Fabrica*, was printed in 1543. In this epochal work Vesalius deployed all his scientific, humanistic, and aesthetic gifts. The *Fabrica* was a more extensive and accurate description of the human body than any put forward by his predecessors: it gave anatomy a new language, and, in the elegance of its printing and organization, a perfection hitherto unknown.

Vesalius's work represented the culmination of the humanistic revival of ancient learning, the introduction of human dissections into medical curricula, and the growth of a European anatomical literature. He performed his dissections with an unprecedented thoroughness. After Vesalius, anatomy became a scientific discipline, with far-reaching implications not only for physiology but also for all of biology. During his own lifetime, however, Vesalius found it easier to correct points of Galenic anatomy than to challenge his physiological framework.

Conflicting reports obscure the final days of Vesalius' life. Apparently he became ill aboard ship while returning to Europe from his pilgrimage. He was put ashore on the Greek island of Zacynthus, where he died.

TYCHO BRAHE (1546–1601)

Danish astronomer whose work in developing astronomical instruments and in measuring and fixing the positions of stars paved the way for future discoveries.

Tycho's father was a privy councillor and later governor of the castle of Helsingborg, in Sweden. His wealthy and childless uncle abducted Tycho at a very early age and, after the initial parental shock was overcome, raised him at his castle in Tostrup, Scania. He also financed the youth's education, which began with the study of law at the University of Copenhagen in 1559–62.

Several important natural events turned Tycho from law to astronomy. The first was the eclipse of the sun predicted for 21 August 1560. Such a prediction seemed audacious and marvellous to a 14-year-old student. His subsequent student life was divided between his daytime lectures on jurisprudence, in response to the wishes of his uncle, and his night-time vigil of the stars. His professor of mathematics helped him with the only printed astronomical book available, the *Almagest* of Ptolemy. Other teachers assisted him in constructing small globes, on which star positions could be plotted, and compasses and cross-staffs, with which he could estimate the angular separation of stars.

Another significant event in Tycho's life occurred in August 1563, when he made his first recorded observation, a conjunction (close approach) of Jupiter and Saturn. He found that the existing almanacs were grossly inaccurate: the tables that described the planetary positions were several days off in predicting this event. In his youthful enthusiasm Tycho decided to devote his life to the accumulation of accurate observations of the heavens, in order to correct the existing tables.

Between 1565 and 1570 or 1572 he travelled widely throughout Europe, studying at Wittenberg, Rostock, Basel, and Augsburg, and acquiring mathematical and astronomical instruments, including a huge quadrant. Tycho then settled in Scania in around 1571 and, on a property owned by a relative, constructed a small observatory. Here occurred the third and most important astronomical event in Tycho's life. On 11 November 1572 he suddenly saw a "new star", brighter than Venus and where no star was supposed to be, in the constellation Cassiopeia. He carefully observed the new star and showed that it lay beyond the moon and therefore was in the realm of the fixed stars.

To the world at that time this was a disquieting discovery. The news that a star could change as dramatically as that described by Tycho, together with the reports of the Copernican theory that the sun, not the Earth, was the centre of the universe, shook confidence in the immutable laws of antiquity and suggested that the chaos and imperfections of Earth were reflected in the heavens. Tycho's discovery of the new star in Cassiopeia and his publication of his observations of it in *De nova stella* ("On the New Star") in 1573 marked his transformation from a Danish dilettante to an astronomer with a European reputation.

Tycho's discovery of the new star had caused to rededicate himself to astronomy, and one immediate decision was to establish a large observatory for regular surveillance of celestial events. His plan to establish this observatory in Germany prompted King Frederick II to keep him in Denmark by granting him title, in 1576, to the island of Ven (formerly Hven), in the middle of The Sound and about halfway between Copenhagen and Helsingør, together with financial support for the observatory and laboratory buildings. Tycho called the observatory Uraniborg, after Urania, the muse of astronomy. Surrounded by scholars and visited by learned travellers from all over Europe, Tycho and his assistants collected observations and substantially corrected nearly every known astronomical record. His observations – the most accurate possible before the invention of the telescope – included a comprehensive study of the solar system and led to the accurate positioning of more than 777 "fixed stars".

Frederick died in 1588, however. Under his son, Christian IV, Tycho's influence dwindled and he eventually left Ven in 1597. After short stays at Rostock and at Wandsbek, near Hamburg, he settled in Prague in 1599 under the patronage of Emperor Rudolf II, who also in later years supported the astronomer Johannes Kepler.

The major portion of Tycho's lifework – making and recording accurate astronomical observations – had already been done at Uraniborg. He attempted to continue his observations at Prague with the few instruments he had salvaged from Uraniborg, but the spirit was not there, and he died in

1601, leaving all his observational data to Kepler, his pupil and assistant in the final years. With these data Kepler laid the groundwork for the work of Sir Isaac Newton.

GIORDANO BRUNO (1548–1600)

Italian philosopher, astronomer, and mathematician, whose theories anticipated modern science.

Bruno was the son of a professional soldier. He was named Filippo at his baptism and was later called "il Nolano" after the place of his birth. In 1562 he went to Naples to study the humanities, logic, and dialectics (argumentation). In 1565 he entered the Dominican convent of San Domenico Maggiore in Naples and assumed the name Giordano. Because of his unorthodox attitudes he was soon suspected of heresy, but nevertheless, in 1572, was ordained as a priest.

In July 1575 Bruno completed a prescribed course on theology. However, he had read two forbidden commentaries by Erasmus and freely discussed the Arian heresy, which denied the divinity of Christ. As a result, a trial for heresy was prepared against him by the provincial father of the order, and he fled to Rome in February 1576. There he found himself unjustly accused of a murder. A second excommunication process was started, and in April 1576 he fled again. He abandoned the Dominican Order, and, after wandering in northern Italy, went in 1578 to Geneva, where he earned his living by proofreading. Finally in 1581, after a brief flirtation with Calvinism, he travelled to Paris and later to England.

At Oxford Bruno started a series of lectures in which he expounded the Copernican theory maintaining the reality of the movement of the Earth. Because of the hostile reception of the Oxonians, however, he went back to London as the guest of the French ambassador. In February 1584 he was invited by Fulke Greville, an English courtier, to discuss his theory of the movement of the Earth with some Oxonian doctors, but the discussion degenerated into a quarrel. A few days later Bruno started writing his Italian dialogues, which constitute the first systematic exposition of his philosophy.

There are six dialogues, three of which are cosmological – on the theory of the universe. In the *Cena de le Ceneri* ("The Ash Wednesday Supper"; 1584), he not only reaffirmed the reality of the heliocentric theory of the heavens but also suggested that the universe is infinite, constituted of innumerable worlds substantially similar to those of the solar system. In the same dialogue he maintained that the Bible should be followed for its moral teaching but not for its astronomical implications. In the *De la causa, principio e uno* ("Concerning the Cause, Principle, and One"; 1584) he elaborated the physical theory on which his conception of the universe was based: "form" and "matter" are intimately united and constitute the "one". Thus, the traditional dualism of the Aristotelian physics was reduced by Bruno to a monistic conception of the world, implying the basic unity of all substances and the coincidence of opposites in the infinite unity of Being. In the *De l'infinito universo e mondi* ("On the Infinite Universe and Worlds"; 1584), he developed his cosmological theory by systematically criticizing Aristotelian physics; he also formulated his Averroistic view of the relationship between philosophy and religion, according to which religion is considered as a means to instruct and govern ignorant people; philosophy as the discipline of the elect who are able to behave themselves and govern others.

In 1585 Bruno returned to Paris. Henry III had abrogated the edict of pacification with the Protestants. Far from adopting a cautious line of behaviour, however, Bruno entered into a polemic with a protégé of the Catholic party, the mathematician Fabrizio Mordente, whom he ridiculed in four *Dialogi*, and in May 1586 he dared to attack Aristotle publicly. Disavowed by former allies, Bruno went to Germany, where he wandered from one university city to another, lecturing and publishing a variety of minor works. These included the *Articuli centum et sexaginta* ("160 Articles"; 1588) against contemporary mathematicians and philosophers, in which he expounded his conception of religion. He also wrote three Latin poems, which re-elaborate the theories expounded in the Italian dialogues and develop Bruno's concept of an atomic basis for matter and being. To publish these he went to Frankfurt am Main, where the senate rejected his application to stay. Nevertheless, he took up residence in the Carmelite convent, lecturing to Protestant doctors and acquiring a reputation of being a "universal man" who, the Prior thought, "did not possess a trace of religion" and who "was chiefly occupied in writing and in the vain and chimerical imagining of novelties."

In 1591, at the invitation of the Venetian patrician Giovanni Mocenigo, Bruno made the fatal move of returning to Italy. As the guest of Mocenigo he took part in the discussions of progressive Venetian aristocrats who, like Bruno, favoured philosophical investigation irrespective of its theological implications. Bruno's liberty came to an end when Mocenigo denounced him to the Venetian Inquisition for his heretical theories. Bruno was arrested and tried. Then, however, the Roman Inquisition demanded his extradition, and on 27 January 1593, Bruno entered the jail of the Roman palace of the Sant'Uffizio (Holy Office).

During the seven-year Roman period of the trial, Bruno at first disclaimed any particular interest in theological matters and reaffirmed the philosophical character of his speculation, but the inquisitors demanded an unconditional retraction of his theories. Bruno then made a desperate attempt to demonstrate that his views were not incompatible with the Christian conception of God and creation, and finally declared that he had nothing to retract. Pope Clement VIII then ordered that he be sentenced as an impenitent and pertinacious heretic. On 8 February 1600, when the death sentence was formally read to him, Bruno addressed his judges, saying: "Perhaps your fear in passing judgment on me is greater than mine in receiving it." Not long after, he was brought to the Campo de' Fiori and burned alive.

Bruno's theories influenced seventeenth-century scientific and philosophical thought and, since the eighteenth century, have been absorbed by many modern philosophers. As a symbol of the freedom of thought, Bruno inspired the European liberal movements of the nineteenth century, particularly the Italian Risorgimento (the movement for national political unity). Because of the variety of his interests, modern scholars are divided as to the chief significance of his work. Bruno's cosmological vision certainly anticipates some fundamental aspects of the modern conception of the universe; his ethical ideas, in contrast with religious ascetical ethics, appeal to modern humanistic activism; and his ideal of religious and philosophical tolerance has influenced liberal thinkers. On the other hand, his emphasis on the magical and the occult has been a source of criticism, as has his impetuous personality. Bruno stands, however, as one of the important figures in the history of Western thought, a precursor of modern civilization.

FRANCIS BACON, VISCOUNT ST ALBAN (1561–1626)

Lawyer, statesman, philosopher, and master of the English tongue, who advocated a new method of acquiring natural knowledge.

Bacon was born at York House off the Strand, London, the younger of the two sons of the lord keeper, Sir Nicholas Bacon, by his second marriage. From 1573 to 1575 he was educated at Trinity College, University of Cambridge, and his distaste for what he termed "unfruitful" Aristotelian philosophy began there.

In 1579 he took up residence at Gray's Inn, an institution for legal education in London. After becoming a barrister in 1582 he progressed through the posts of Reader (lecturer at the Inn), Bencher (senior member of the Inn), and Queen's (from 1603 King's) Counsel extraordinary to those of Solicitor General and Attorney General. Even as successful a legal career as this, however, did not satisfy his political and philosophical ambitions.

When Elizabeth I died in 1603, Bacon's letter-writing ability was directed to finding a place for himself and a use for his talents in James I's services. Through the influence of his cousin Robert Cecil, Bacon was one of the 300 new knights dubbed in 1603. The following year he was confirmed as Learned Counsel and sat in the first Parliament of the new reign in the debates of its first session.

In the autumn of 1605 he published his *Advancement of Learning*, dedicated to the King. Preferment in the royal service, however, still eluded him, and it was not until June 1607 that his petitions and his vigorous, though vain, efforts to persuade the Commons to accept the King's proposals for union with Scotland were rewarded with the post of solicitor general. Even then, his political influence remained negligible. In 1609 his *De Sapientia Veterum* ("The Wisdom of the Ancients"), in which he expounded what he took to be the hidden practical meaning embodied in ancient myths, was published – it proved to be, after the *Essayes* (1597; a collection of writings on every aspect of life), his most popular book in his own lifetime. *The New Atlantis*, his far-seeing scientific utopian work, appears to have been written in 1614, but did not get into print until 1626.

After Salisbury's death in 1612 Bacon renewed his efforts to gain influence with the King, yet nothing was forthcoming until March 1617, when he was named as Lord Keeper of the Great Seal. In 1618 he was made Lord Chancellor and Baron Verulam, and in 1620/21 he was created Viscount St Alban. The main reason for this progress was his unsparing service in Parliament and the court, together with persistent letters of self-recommendation; according to the traditional account, however, he was also aided by his association with George Villiers, later Duke of Buckingham, the King's new favourite.

By 1621 Bacon must have seemed impregnable, a favourite not by charm (though he was witty and had a dry sense of humour) but by sheer usefulness and loyalty to his sovereign; lavish in public expenditure (he was once the sole provider of a court masque); dignified in his affluence and liberal in his household. He won the attention of scholars abroad as the author of the *Novum Organum* ("New Instrument"), published in 1620, and the developer of the *Instauratio Magna* ("Great Instauration"), a comprehensive plan to reorganize the sciences and to restore man to that mastery over nature that he was thought to have lost by the fall of Adam.

Bacon's ambitious scheme for *Instauratio Magna* was never completed. Its first part, *De Augmentis Scientiarum* ("On the Advancement of Learning"), appeared in 1623 and is an expanded, Latinized version of his earlier work the *Advancement of Learning*, published in 1605 – the first really important philosophical book to be written in English. The *De Augmentis Scientiarum* contains a division of the sciences, a project that had not been embarked on to any great purpose since Aristotle or, in a smaller way, since the Stoics.

The second part of Bacon's scheme, the *Novum Organum*, which had already appeared in 1620, gives "true directions concerning the interpretation of nature" – in other words, an account of the correct method of acquiring natural knowledge. Bacon believed this to be his most important contribution to science, and this is the body of ideas with which his name is most closely associated. The fields of possible knowledge having been charted in *De Augmentis Scientiarum*, the proper method for their cultivation was set out in *Novum Organum*.

The core of Bacon's philosophy of science is the account of inductive reasoning given in Book II of *Novum Organum*. The defect of all previous systems of belief about nature, he argued, lay in the inadequate treatment of the general propositions from which the deductions were made. Either they were the result of precipitate generalization from one or two cases, or they were uncritically assumed to be self-evident on the basis of their familiarity and general acceptance.

The crucial point, Bacon realized, is that induction must work by elimination – not, as it does in common life and the defective scientific tradition, by simple enumeration. He devised tables, or formal devices for the presentation of singular pieces of evidence, in order to facilitate the rapid discovery of false generalizations. What survives this eliminative screening, Bacon assumed, may be taken to be true. An exemplary demonstration of this method in *Novum Organum* is the application of Bacon's inductive "tables" to show heat to be a kind of motion of particles.

The conception of a scientific research establishment, which Bacon developed in his utopia, *The New Atlantis*, may be a more important contribution to science than his theory of induction. Here the idea of science as a collaborative undertaking, conducted in an impersonally methodical fashion and animated by the intention to give material benefits to mankind, is set out with literary force.

Bacon acknowledges something he calls first philosophy, which is secular but not confined to nature or to society. It is concerned with the principles, such as they are, that are common to all the sciences. Natural philosophy divides on the one hand into natural science as theory, and on the other into the practical discipline of applying natural science's findings to improving the human condition, or "the relief of man's estate" – which he misleadingly describes as "natural magic". The former is "the inquisition of causes"; the latter the "the production of effects". To subdivide still further, Bacon considers natural science to be made up of physics and metaphysics. Physics, in his interpretation, is the science of observable correlations; metaphysics is the more theoretical science of the underlying structural factors that explain observable regularities. Each has its practical, or technological, partner: that of physics is mechanics; that of metaphysics, natural magic. It is to the latter that one must look for the real transformation of the human condition through scientific progress. Mechanics is just levers and pulleys.

While Bacon the philosopher was developing his ideas, his political enemies were plotting. In 1621, two charges of bribery were raised against him before a committee of grievances. He lost all political power and was banished from the court, and was briefly imprisoned in the Tower of London. Despite all this his courage held – the last years of his life were spent in work far more valuable than anything he had accomplished in his high office.

Cut off from other services, he offered his literary powers to provide the King with a digest of the laws, a history of Great Britain, and biographies of Tudor monarchs. He prepared memorandums on usury and on the prospects of a war with Spain, and he expressed views on educational reforms. Bacon in adversity showed patience, unimpaired intellectual vigour, and fortitude. Physical deprivation distressed him, but what hurt most was the loss of favour. It was not until 20 January 1622/23 that he was admitted to kiss the King's hand, but a full pardon never came. Finally, in March 1626, driving one day near Highgate, a district to the north of London, and deciding on impulse to discover whether snow would delay the process of putrefaction, he stopped his carriage, purchased a hen, and stuffed it with snow. He was seized with a sudden chill, which brought on bronchitis, and he died at the Earl of Arundel's house nearby on 9 April 1626.

It has been suggested that Bacon's thought received proper recognition only with nineteenth-century biology, which, unlike mathematical physics, really is Baconian in procedure. Darwin (1809–82) undoubtedly thought so. Bacon's belief that a new science could contribute to "the relief of man's estate" also had to await its time. In the seventeenth century the chief inventions that flowed from science were of instruments that enabled science to progress further. Today Bacon is best known among philosophers as the symbol of the idea – widely held to be mistaken – that science is inductive. Although there is more to his thought than this, this concept is indeed central, and, even if it is inaccurate, it is as well to have it so boldly and magnificently presented.

GALILEO GALILEI (1564–1642)

Italian natural philosopher, astronomer, and mathematician who made fundamental contributions to the sciences of motion and astronomy, and to the development of the scientific method.

Galileo was born in Pisa, Tuscany. He was the oldest son of Vincenzo Galilei, a musician who made important contributions to the theory and practice of music and who may have performed some experiments with Galileo in 1588–9 on the relationship between pitch and the tension of strings. The family moved to Florence in the early 1570s, where the Galilei family had lived for generations. In 1581 Galileo matriculated at the University of Pisa, where he was to study medicine. However, he became enamoured with mathematics and, against the protests of his father, decided to make the mathematical subjects and philosophy his profession. During this period he designed a new form of hydrostatic balance for weighing small quantities and wrote a short treatise, *La bilancetta* ("The Little Balance"), which circulated in manuscript form. He also began his studies on motion, which he pursued steadily for the next two decades.

In 1588 Galileo applied for the chair of mathematics at the University of Bologna but was unsuccessful. His reputation was, however, increasing. He also conceived some ingenious theorems on centres of gravity, again circulated in manuscript, which brought him recognition among mathematicians. As a result, he obtained the chair of mathematics at the University of Pisa in 1589. There, according to his first biographer, Vincenzo Viviani, Galileo demonstrated – by dropping bodies of different weights from the top of the famous Leaning Tower – that the speed of fall of a heavy object is not proportional to its weight, as Aristotle had claimed.

The manuscript tract *De motu* ("On Motion"), finished during this period, shows that Galileo was abandoning Aristotelian notions about motion and was instead taking an Archimedean approach to the problem. By 1609 he had determined that the distance fallen by a body is proportional to the square of the elapsed time (i.e. the law of falling bodies), and that the trajectory of a projectile is a parabola – both conclusions that contradicted Aristotelian physics.

At this point, however, Galileo's career took a dramatic turn. In the spring of 1609 he heard that in the Netherlands an instrument had been invented that showed distant things as though they were nearby. By trial and error, he soon determined the secret of the invention and made his own three-powered spyglass from lenses for sale in spectacle-makers' shops. Others had done the same, but what set Galileo apart was that he quickly calculated how to improve the instrument, taught himself the art of lens grinding, and produced increasingly powerful telescopes. In August of that year he presented an eight-powered instrument to the Venetian Senate. He was rewarded with life tenure and a doubling of his salary. He was now one of the highest-paid professors at the university.

In the autumn of 1609 Galileo began observing the heavens with instruments that magnified up to 20 times. In December he drew the moon's phases as seen through the telescope, showing that the moon's surface is not smooth, as had been thought, but is rough and uneven. In January 1610 he discovered four moons revolving around Jupiter. He also found that the telescope showed many more stars than are visible with the naked eye. He produced a little book, *Sidereus Nuncius* ("The Sidereal Messenger"), in which he described these momentous discoveries.

Galileo was now a courtier and lived the life of a gentleman. Before he left Padua he had discovered the puzzling appearance of Saturn, later to be shown as caused by a ring surrounding it, and in Florence he discovered that Venus goes through phases just as the moon does. Although these discoveries did not prove that the Earth is a planet orbiting the sun, they undermined Aristotelian cosmology: the absolute difference between the corrupt earthly region and the perfect and unchanging heavens was proved wrong by the mountainous surface of the moon; the moons of Jupiter showed that there had to be more than one centre of motion in the universe; and the phases of Venus showed that it (and, by implication, Mercury) revolves around the sun. As a result, Galileo was confirmed in his belief, which he had probably held for decades but which had not been central to his studies, that the sun is the centre of the universe and that the Earth is a planet, as Copernicus had argued. Galileo's conversion to Copernicanism would be a key turning point in the scientific revolution.

Galileo's increasingly overt Copernicanism began to cause trouble for him. In 1613 he wrote a letter to his student Benedetto Castelli in Pisa about the problem of squaring the Copernican theory with certain biblical passages. Inaccurate copies of this letter were sent by Galileo's enemies to the Inquisition in Rome, and he had to retrieve the letter and send an accurate copy. Several Dominican fathers in Florence lodged complaints against Galileo in Rome, and he went to Rome to defend the Copernican cause and his good name. But the tide in Rome was turning against the Copernican theory and he was effectively muzzled on the issue.

He recovered slowly from this setback. Through a student, he entered a controversy about the nature of comets occasioned by the appearance of three comets in 1618. After several exchanges, mainly with Orazio Grassi, a professor of mathematics at the Collegio Romano, he finally entered the argument under his own name. *Il saggiatore* ("The Assayer"), published in 1623, was a brilliant polemic on physical reality and an exposition of the new scientific method. In it, Galileo discussed the method of the newly emerging science, arguing:

Philosophy is written in this grand book, the universe, which stands continually open to our gaze. But the book cannot be understood unless one first learns to comprehend the language and read the letters in which it is composed. It is written in the language of mathematics, and its characters are triangles, circles, and other geometric figures without which it is humanly impossible to understand a single word of it.

Publication of *Il saggiatore* came at an auspicious moment, for Maffeo Cardinal Barberini (1568–1644), a friend, admirer, and patron of Galileo for a decade, was named Pope Urban VIII as the book was going to press. Galileo's friends quickly arranged to have it dedicated to the new pope. In 1624 Galileo went to Rome and had six interviews with Urban VIII. The pope gave Galileo permission to write a book about theories of the universe, but warned him to treat the Copernican theory only hypothetically. The book, *Dialogo sopra i due massimi sistemi del mondo, tolemaico e copernicano* ("Dialogue Concerning the Two Chief World Systems, Ptolemaic and Copernican"), was finished in 1630. Galileo sent it to the Roman censor, who forwarded several serious criticisms of the book to his colleagues in Florence. After writing a preface in which he professed that what followed was written hypothetically, Galileo had little trouble getting the book through the Florentine censors, and it appeared in Florence in 1632.

The reaction against the book was swift. The pope convened a special commission to examine the book and make recommendations, and the commission found that Galileo had not really treated the Copernican theory hypothetically and recommended that a case be brought against him by the Inquisition. Galileo was summoned to Rome in 1633. During his first appearance before the Inquisition, he was confronted with the 1616 edict recording that he was forbidden to discuss the Copernican theory. In his defence, he produced a letter from Cardinal Bellarmine, by then dead, stating that he was admonished only not to hold or defend the theory. The case was at an impasse, and, in what can only be called a plea bargain, Galileo confessed to having overstated his case. He was pronounced to be vehemently suspect of heresy and was condemned to life imprisonment and made to abjure formally. There is no evidence that at this time he whispered, "*Eppur si muove*" ("And yet it moves").

After the process he spent six months at the palace of Ascanio Piccolomini (c. 1590–1671), the archbishop of Siena and a friend and patron, and then moved into a villa near Arcetri, in the hills above Florence. He spent the rest of his life there. His daughter, Sister Maria Celeste, who was in a nearby nunnery, was a great comfort to her father until her untimely death in 1634. Galileo was by then 70 years old, yet he kept working. In Siena he had begun a new book on the sciences of motion and strength of materials. There he wrote up his unpublished studies, which had been interrupted by his interest in the telescope in 1609 and pursued intermittently since. The book was spirited out of Italy and published in Leiden, Netherlands, in 1638 under the title *Discorsi e dimostrazioni matematiche intorno a due nuove scienze attenenti alla meccanica* ("Dialogues Concerning Two New Sciences"). Galileo here treated for the first time the bending and breaking of beams, and summarized his mathematical and experimental investigations of motion, including the law of falling bodies and the parabolic path of projectiles as a result of the mixing of two motions, constant speed and uniform acceleration. By then he had become blind, and he spent his time working with a young student, Vincenzo Viviani, who was with him when he died.

JOHANNES KEPLER (1571–1630)

German astronomer who discovered
three major laws of planetary motion.

Kepler came from a very modest family in a small German town called Weil der Stadt. A ducal scholarship made it possible for him to attend the Lutheran *Stift*, or seminary, at the University of Tübingen, where he began his university studies in 1589. It was expected that the boys who graduated from these schools would go on to become schoolteachers, ministers, or state functionaries. Kepler had planned to become a theologian, but his life did not work out quite as he expected.

At Tübingen Kepler came under the influence of the professor of mathematics, Michael Maestlin, one of the most talented astronomers in Germany. Maestlin had once been a Lutheran pastor; he was also, privately, one of the few adherents of the Copernican theory in the late sixteenth century, although very cautious about expressing his views in print. Kepler quickly grasped the main ideas in Copernicus's work and was tutored in its complex details by Maestlin. He felt that Copernicus had hit upon an account of the universe that contained the mark of divine planning – literally a revelation. Early in the 1590s, while still a student, Kepler would make it his mission to demonstrate rigorously what Copernicus had only guessed to be the case. And he did so in an explicitly religious and philosophical vocabulary.

The ideas that Kepler would pursue for the rest of his life were already present in his first work, *Mysterium cosmographicum* ("Cosmographic Mystery"; 1596). In place of the tradition that individual incorporeal souls push the planets, and instead of Copernicus's passive, resting sun, Kepler posited the hypothesis that a single force from the sun accounts for the increasingly long periods of motion as the planetary distances increase. He did not yet have an exact mathematical description for this relation, but he intuited a connection: the universe is a system of magnetic bodies in which, with corresponding like poles repelling and unlike poles attracting, the rotating sun sweeps the planets around.

But there was something more. In 1600 Tycho Brahe invited Kepler to join his court at Castle Benátky near Prague. When Tycho died suddenly a year later, Kepler quickly succeeded him as imperial mathematician to Holy Roman Emperor Rudolf II (ruled 1576–1612). Kepler's first publication as imperial mathematician was a work that broke with the theoretical principles of Ptolemaic astrology. Called *De Fundamentis Astrologiae Certioribus* ("Concerning the More Certain Fundamentals of Astrology"; 1601), this work proposed to make astrology "more certain" by basing it on new physical and harmonic principles. It showed both the importance of astrological practice at the imperial court and Kepler's intellectual independence in rejecting much of what was claimed to be known about stellar influence.

The relatively great intellectual freedom possible at Rudolf's court was now augmented by Kepler's unexpected inheritance of a critical resource: Tycho's observations. In his lifetime Tycho had been ungenerous in sharing his work. After his death, although there was a political struggle with Tycho's heirs, Kepler was ultimately able to work with data accurate to within 2 minutes (one-thirtieth of one degree) of arc. Without data of such precision to back up his solar hypothesis, Kepler would have been unable to discover his "first law" (1605), that Mars moves in an elliptical orbit with the sun at one focus (one of the two points, each called a focus, that define the shape of an ellipse).

He subsequently determined that the time necessary to traverse any arc of a planetary orbit is proportional to the area of the sector between the central body and that arc (the "area law"), and that there is an exact relationship between the squares of the planets' periodic times and the cubes of the radii of their orbits (the "harmonic law"). Kepler himself did not call these discoveries "laws", as would become customary after Isaac Newton (1642–1727) derived them from a new and quite different set of general physical principles. He regarded them as celestial harmonies that reflected God's design for the universe.

During the creative burst of the early Prague period (1601–05) Kepler also wrote important treatises on the nature of light and on the sudden appearance of a new star (*De Stella Nova*, "On the New Star"; 1606). Kepler first noticed the star – now known to have been a supernova – in October 1604, not long after a conjunction of Jupiter and Saturn in 1603. The astrological importance of the long-awaited conjunction (such configurations take place every 20 years) was heightened by the unexpected appearance of the supernova. Typically, Kepler used the occasion both to render practical predictions (e.g., the collapse of Islam and the return of Christ) and to speculate theoretically about the universe.

After Galileo built a telescope in 1609 and announced hitherto-unknown objects in the heavens (e.g., moons revolving around Jupiter) and imperfections of the lunar surface, he sent Kepler his account in *Siderius Nuncius* ("The Sidereal Messenger"; 1610). Kepler responded with three important treatises. The first was his *Dissertatio cum Nuncio Sidereo* ("Conversation with the Sidereal Messenger"; 1610), in which, among other things, he speculated that the distances of the newly discovered Jovian moons might agree with the ratios of the rhombic dodecahedron, triacontahedron, and cube. The second was a theoretical work on the optics of the telescope, *Dioptrice* ("Dioptrics"; 1611), including a description of a new type of telescope using two convex lenses. The third was based upon his own observations of Jupiter, made between 30 August and 9 September 1610, and published as *Narratio de Jovis Satellitibus* ("Narration Concerning the Jovian Satellites"; 1611). These works provided strong support for Galileo's discoveries, and Galileo, who had never been especially generous to Kepler, wrote to him, "I thank you because you were the first one, and practically the only one, to have complete faith in my assertions."

In 1611 Kepler's life took a turn for the worse. His wife, Barbara, became ill, and his three children contracted smallpox; one of his sons died. Shortly afterwards, Emperor Rudolf abdicated his throne. Although Kepler hoped to return to an academic post at Tübingen, there was resistance from the theology faculty. Meanwhile, he was appointed to the position (created for him) of district mathematician in Linz. He continued to hold the position of imperial mathematician under the new emperor, Matthias, although he was physically removed from the court in Prague. Kepler stayed in Linz until 1626.

The Linz authorities had expected that Kepler would use most of his time to work on and complete the astronomical tables begun by Tycho. But the work was tedious, and Kepler continued his search for the world harmonies that had inspired him since his youth. In 1619 his *Harmonice Mundi* ("Harmonies of the World") brought together more than two decades of investigations into the archetypal principles of the world: geometrical, musical, metaphysical, astrological, astronomical, and those principles pertaining to the soul. Finally, Kepler published the first textbook of Copernican astronomy, *Epitome Astronomiae Copernicanae* ("Epitome of Copernican Astronomy"; 1618–21). The title mimicked Maestlin's traditional-style textbook, but the content could not have been more different.

The *Epitome* began with the elements of astronomy but then gathered together all the arguments for Copernicus's theory and added to them Kepler's harmonics and new rules of planetary motion. This work would prove to be the most important theoretical resource for the Copernicans in the seventeenth century. It was capped by the appearance of *Tabulae Rudolphinae* ("Rudolphine Tables"; 1627). The *Epitome* and the *Tabulae Rudolphinae* cast heliostatic astronomy and astrology into a form where detailed and extensive counter-argument would force opponents to engage with its claims or silently ignore them to their disadvantage. Eventually Newton would simply take over Kepler's laws while ignoring all reference to their original theological and philosophical framework.

In 1627 Kepler found a new patron in the Imperial General Albrecht von Wallenstein, who sent him to Sagan in Silesia and supported the construction of a printing press for him. In return Wallenstein expected horoscopes from Kepler – who accurately predicted "horrible disorders" for March 1634, close to the actual date of Wallenstein's murder on 25 February 1634. Kepler was less successful in his ever-continuing struggle to collect monies owed him. In August 1630 Wallenstein lost his position as commander in chief, and in October Kepler left for Regensburg in the hope of collecting interest on some Austrian bonds. But soon after arriving he became seriously ill with fever, and on 15 November he died.

WILLIAM HARVEY (1578–1657)

English physician and discoverer of the true
nature of the circulation of the blood and
of the function of the heart as a pump.

Little is known of Harvey's boyhood in the countryside of Kent. During the years 1588 to 1593 he was at the King's School attached to the cathedral at Canterbury. In his 16th year Harvey entered Gonville and Caius College, University of Cambridge, where he was awarded a scholarship in 1593. Although he attended Caius College because of its special interest in educating doctors, his training was grossly inadequate. He was absent from the university for the greater part of his last year (1598–9) because of illness – probably malaria – but had received the BA degree in 1597. Determined to continue with medical training, he began a two-and-a-half-year course of study at the University of Padua, reputed to have the best medical school in Europe. His teacher, Hieronymus Fabricius ab Aquapendente, was a celebrated anatomist, and it was in the now-famous oval Anatomy Theatre at the university that Harvey first recognized the problems posed by the function of the beating heart and the properties of the blood passing through it.

His 28 months at Padua are only meagrely documented, but it is clear that he was outstanding among the students of his year.

After receiving his diploma as Doctor of Medicine of Padua in April 1602, he returned to England. In 1607 he obtained a fellowship of the College of Physicians, which entitled him to seek an appointment as physician to one of the two great hospitals then serving London – St Bartholomew's and St Thomas's. In 1609 the King gave Harvey a recommendation for an appointment at St Bartholomew's, which was conveniently near his house in St Martin's. He was given the post of assistant physician, and, when the physician died in the summer of that year, Harvey succeeded him. Harvey held this office for 34 years, until 1643 when he was displaced for political reasons by Oliver Cromwell's party, then in power in London.

These years saw the development and culmination of his active career as physician and scientific innovator. He developed a large private practice, attending many of the most distinguished citizens and, in about 1618, was made physician extraordinary to King James I. There can be no doubt that Harvey was for many years one of the most widely trusted doctors in England, although his unorthodox views on the circulation of the blood did subsequently injure his practice.

In 1628 he finally published *Exercitatio Anatomica de Motu Cordis et Sanguinis in Animalibus* ("An Anatomical Exercise Concerning the Motion of the Heart and Blood in Animals"). The volume established the true nature of the circulation of the blood, which had been previously largely misunderstood. He disposed of the idea that the blood vessels contained air, elucidated the function of the valves in the heart in maintaining the flow of blood in one direction only when the ventricles contracted, and proved that the arterial pulse resulted from the passive filling of the arteries by the contraction of the heart and not by active contraction of their walls.

Harvey's book made him famous throughout Europe, although the overthrow of time-hallowed beliefs attracted virulent attacks and much abuse. He refused to indulge in controversy and made no reply until 1649, when he published a small book answering the criticisms of a French anatomist, Jean Riolan. In this work he reiterated some of his former arguments and utterly demolished Riolan's objections.

At the start of the Civil War in 1642, Harvey was with the King and was in charge of the two princes, Charles and James. When the defeated King fled from Oxford to surrender himself to the Scots, Harvey joined him for a time at Newcastle but was forced to leave the King when he was handed over to the parliamentary army, and was not allowed to go to him when he was imprisoned in the Isle of Wight. Harvey had never been much interested in politics but felt a deep personal regard for the King and, after his execution in 1649, was a broken and unhappy man.

Two years later, however, he published his second great book. After the publication of *De Motu Cordis* he had continued active research into the difficult subject of reproduction in animals. This led in 1651 to the publication of *Exercitationes de Generatione Animalium* ("Anatomical Exercitations Concerning the Generation of Animals") through the persuasions of his younger friend Sir George Ent, a fellow of the college. The book is mainly concerned with the development of the chick in hens' eggs, and Harvey insisted throughout that in all living things the origin of the embryo is to be found in the egg.

In his last years, under Cromwell's Protectorate, Harvey was regarded as a political "delinquent" owing to his long association with King Charles and was forced to spend most of his time lodging in one or another of his brothers' houses outside London. Though he corresponded with many distinguished foreign doctors he was reluctant to engage in any further scientific research, saw few patients, and took little part in the affairs of the College of Physicians.

RENÉ DESCARTES (1596–1650)

French mathematician, scientist, and philosopher,
who applied an original system of methodical
doubt famed for his *Discourse on Method*
and the dictum "I think therefore I am."

Descartes was born in La Haye (now Descartes), France. In
1606 he was sent to the Jesuit college at La Flèche where 1,200
young men were trained for careers in military engineering,
the judiciary, and government administration. In addition
to classical studies, science, mathematics, and metaphysics –
Aristotle was taught from scholastic commentaries – they
studied acting, music, poetry, dancing, riding, and fencing.
In 1614 Descartes went to Poitiers, where he took a law degree
in 1616, and then travelled in the Netherlands, where he spent
15 months as an informal student of mathematics and military
architecture in the peacetime army of the Protestant stad-
holder, Prince Maurice (ruled 1585–1625).

Descartes spent the period 1619 to 1628 travelling in northern and southern Europe, where, as he later explained, he
studied "the book of the world". While in Bohemia in 1619 he
invented analytic geometry, a method of solving geometric
problems algebraically and algebraic problems geometrically.
He also devised a universal method of deductive reasoning,
based on mathematics, that is applicable to all the sciences.
This method, which he later formulated in *Discourse on
Method* (1637) and *Rules for the Direction of the Mind*
(written by 1628 but not published until 1701), consists of
four rules: (1) accept nothing as true that is not self-evident, (2)
divide problems into their simplest parts, (3) solve problems by
proceeding from simple to complex, and (4) recheck the
reasoning. These rules are a direct application of mathematical

procedures. In addition, he insisted that all key notions and the limits of each problem must be clearly defined.

Descartes moved to Paris in 1622 and enjoyed a life of leisure there. He befriended the mathematician Claude Mydorge and Father Marin Mersenne, a man of universal learning who corresponded with hundreds of scholars, writers, mathematicians, and scientists and who became Descartes's main contact with the larger intellectual world. During this time Descartes regularly hid from his friends to work, writing treatises, now lost, on fencing and metals. He then turned once more to the Netherlands – a haven of tolerance where he could be an original, independent thinker without fear of being burned at the stake or being drafted into the armies then prosecuting the Catholic Counter-Reformation.

Descartes's *Discourse on Method* is one of the first important modern philosophical works not written in Latin. Descartes said that he wrote in French so that all who had good sense, including women, could read his work and learn to think for themselves. He believed that everyone could tell true from false by the natural light of reason. In three essays accompanying the *Discourse*, he illustrated his method for utilizing reason in the search for truth in the sciences: in *Dioptrics* he derived the law of refraction, in *Meteorology* he explained the rainbow, and in *Geometry* he gave an exposition of his analytic geometry. In the *Discourse* he also provided a provisional moral code (later presented as final) for use while seeking truth: (1) obey local customs and laws, (2) make decisions on the best evidence and then stick to them firmly as though they were certain, (3) change desires rather than the world, and (4) always seek truth. This code exhibits Descartes's prudential conservatism, decisiveness, stoicism, and dedication. The *Discourse* and other works illustrate Descartes's conception of knowledge as being like a tree – in its interconnectedness and in the grounding provided to higher forms of knowledge by lower or more fundamental ones. Thus, for Descartes, metaphysics corresponds to the roots of the tree; physics to the trunk; and medicine, mechanics, and morals to the branches.

In 1641 Descartes published the *Meditations on First Philosophy, in Which is Proved the Existence of God and the Immortality of the Soul.* The *Meditations* is characterized by Descartes's use of methodic doubt, a systematic procedure of rejecting as though false all types of belief in which one has ever been, or could ever be, deceived. Thus, Descartes's apparent knowledge based on authority is set aside, because even experts are sometimes wrong. His beliefs from sensory experience are declared untrustworthy, because such experience is sometimes misleading, as when a square tower appears round from a distance. Even his beliefs about the objects in his immediate vicinity may be mistaken, because, as he notes, he often has dreams about objects that do not exist, and he has no way of knowing with certainty whether he is dreaming or awake. Finally, his apparent knowledge of simple and general truths of reasoning that do not depend on sense experience – such as "2 + 3 = 5" or "a square has four sides" – is also unreliable, because God could have made him in such a way that, for example, he goes wrong every time he counts. As a way of summarizing the universal doubt into which he has fallen, Descartes supposes that an "evil genius of the utmost power and cunning has employed all his energies in order to deceive me." Although at this stage there is seemingly no belief about which he cannot entertain doubt, Descartes finds certainty in the intuition that, when he is thinking – even if he is being deceived – he must exist. In the *Dis-*

course, Descartes expresses this intuition in the dictum "I think, therefore I am"; but because "therefore" suggests that the intuition is an argument – though it is not – in the *Meditations* he says merely, "I think, I am" ("*Cogito, sum*"). Descartes also advances a proof for the existence of God. He begins with the proposition that he has an innate idea of God as a perfect being and then concludes that God necessarily exists, because, if he did not, he would not be perfect. This ontological argument for God's existence, originally due to the English logician St Anselm of Canterbury (*c.* 1033–1109), is at the heart of Descartes's rationalism, for it establishes certain knowledge about an existing thing solely on the basis of reasoning from innate ideas, with no help from sensory experience.

In 1644 Descartes published *Principles of Philosophy*, a compilation of his physics and metaphysics in which he argues that human beings can be conditioned by experience to have specific emotional responses. This insight is the basis of Descartes's defence of free will and of the mind's ability to control the body.

Descartes's translator's brother-in-law, Hector Pierre Chanut, who was a French resident in Sweden and later ambassador, helped to procure a pension for Descartes from Louis XIV, though it was never paid. Later, Chanut engineered an invitation for Descartes to the court of Queen Christina, who by the close of the Thirty Years' War (1618–48) had become one of the most important and powerful monarchs in Europe. Descartes went reluctantly, maybe because he needed patronage, arriving early in October 1649.

In Sweden – where, Descartes said, in winter men's thoughts freeze like the water – the 22-year-old Christina perversely made the 53-year-old Descartes rise before 5 a.m. to give her philosophy lessons, even though she knew of his habit of lying in bed until 11 am. While delivering these statutes to the queen at 5 a.m. on 1 February 1650, he caught a chill and soon developed pneumonia. He died in Stockholm on February 11. Many pious last words have been attributed to him, but the most trustworthy report is that of his German valet, who said that Descartes was in a coma and died without saying anything at all.

ROBERT BOYLE (1627–1691) AND ROBERT HOOKE (1635–1703)

British natural philosophers who worked together on a number of early chemical experiments.

Boyle was born into one of the wealthiest families in Britain. He was the 14th child of Richard Boyle, the first Earl of Cork, by his second wife, Catherine, daughter of Sir Geoffrey Fenton, Secretary of State for Ireland. At the age of eight Boyle began his formal education at Eton College, where his studious nature quickly became apparent, and in 1639 he and his brother Francis embarked on a grand tour of the continent with their tutor Isaac Marcombes. In 1649, after his return to England, Boyle began investigating nature via scientific experimentation, a process that enthralled him.

Boyle spent much of 1652–4 in Ireland overseeing his hereditary lands, and he also performed some anatomic dissections. He took up residence at Oxford from c. 1656 until 1668. There he was exposed to the latest developments in natural philosophy and became associated with a group of notable natural philosophers and physicians, including John Wilkins, Christopher Wren, and John

Locke. These individuals, together with a few others, formed the "Experimental Philosophy Club", which at times convened in Boyle's lodgings. Much of Boyle's best-known work dates from this period.

In 1659 Boyle and Robert Hooke, a native of the Isle of Wight, England, completed the construction of their famous air pump and used it to study pneumatics. Their resultant discoveries regarding air pressure and the vacuum appeared in Boyle's first scientific publication, *New Experiments Physico-Mechanicall, Touching the Spring of the Air and its Effects* (1660). Boyle and Hooke discovered several physical characteristics of air, including its role in combustion, respiration, and the transmission of sound. One of their findings, published in 1662, later became known as "Boyle's law". It expresses the inverse relationship that exists between the pressure and volume of a gas, and was determined by measuring the volume occupied by a constant quantity of air when compressed by differing weights of mercury.

Boyle's scientific work is characterized by its reliance on experiment and observation and his reluctance to formulate generalized theories. He advocated a "mechanical philosophy" that saw the universe as a huge machine or clock in which all natural phenomena were accountable purely by mechanical, clockwork motion. His contributions to chemistry were based on a mechanical "corpuscularian hypothesis" – a brand of atomism which claimed that everything was composed of minute (but not indivisible) particles of a single universal matter and that these particles were only differentiable by their shape and motion. Among his most influential writings were *The Sceptical Chymist* (1661) and the *Origine of Formes and Qualities* (1666), which used chemical phenomena to support the corpuscularian hypothesis. Overall, Boyle argued so strongly for the need to apply the principles and methods of chemistry to the study of the natural world and to medicine that he later gained the appellation of the "father of chemistry".

In 1662, after leaving Boyle's service, Robert Hooke was appointed curator of experiments to the Royal Society and was elected a fellow the following year. Here he performed myriad experiments for the debating club: he was one of the first men to build a Gregorian reflecting telescope; he discovered the fifth star in the Trapezium, an asterism in the constellation Orion, in 1664; and he first suggested that Jupiter rotates on its axis. His detailed sketches of Mars were used in the nineteenth century to determine that planet's rate of rotation. In 1665 he was appointed professor of geometry in Gresham College.

In *Micrographia* ("Small Drawings"; 1665) Hooke included his studies and illustrations of the crystal structure of snowflakes, discussed the possibility of manufacturing artificial fibres by a process similar to the spinning of the silkworm, and first used the word "cell" to name the microscopic honeycomb cavities in cork. His studies of microscopic fossils led him to become one of the first proponents of a theory of evolution.

Hooke suggested that the force of gravity could be measured by using the motion of a pendulum (1666), and attempted to show that the Earth and moon follow an elliptical path around the sun. In 1672 he discovered the phenomenon of diffraction (the bending of light rays around corners): to explain it, he offered the wave theory of light. He stated the inverse square law to describe planetary motions in 1678 – a law that Newton (1642–1727) later used in modified form. Hooke complained that he was not given sufficient credit for the law and became involved in bitter controversy with Newton. Hooke was the first man to state in general that all matter expands when heated and that air is made up of particles separated from each other by relatively large distances. He also discovered the law of elasticity, known as "Hooke's law", which states that the stretching of a solid body (e.g., metal or wood) is proportional to the force applied to it. This law laid the basis for studies of stress and strain and for the understanding of elastic materials.

In 1668 Boyle left Oxford and took up residence with his sister Katherine Jones, Vicountess Ranelagh, in her house on Pall Mall in London. There he set up an active laboratory, employed assistants, received visitors, and published at least one book nearly every year. Living in London also provided him the opportunity to participate actively in the Royal Society.

Throughout his adult life, Boyle was sickly, suffering from weak eyes and hands, recurring illnesses, and one or more strokes. He died at the age of 64 after a short illness exacerbated by his grief over Katherine's death a week earlier. He left his papers to the Royal Society and a bequest for establishing a series of lectures, which became known as the Boyle Lectures, in defence of Christianity.

JOHN RAY (1627–1705)

English naturalist and botanist who contributed significantly to progress in taxonomy.

Ray was the son of the village blacksmith in Black Notley, Essex, and attended the grammar school in nearby Braintree. In 1644, with the aid of a fund that had been left in trust to support needy scholars at the University of Cambridge, he matriculated at St Catherine's Hall there, and moved to Trinity

College in 1646. Ray had come to Cambridge at the right time for one with his talents, for he found a circle of friends with whom he pursued anatomical and chemical studies. He was elected to a fellowship at Trinity the following year, and during the next 13 years he lived quietly in his collegiate cloister.

Ray's string of fortunate circumstances ended with the Restoration. Although he was never an excited partisan, he was thoroughly Puritan in spirit and refused to take the oath that was prescribed by the Act of Uniformity. In 1662 he lost his fellowship. Prosperous friends supported him during the subsequent 43 years while he pursued his career as a naturalist. That career had already begun with the publication of his first work in 1660, a catalogue of plants growing around Cambridge. After he had exhausted the Cambridge area as a subject for his studies, Ray began to explore the rest of Britain. An expedition in 1662 to Wales and Cornwall with the naturalist Francis Willughby was a turning point in his life. Willughby and Ray agreed to undertake a study of the complete natural history of living things, with Ray responsible for the plant kingdom and Willughby the animal.

The first fruit of the agreement, a tour of the European continent lasting from 1663 to 1666, greatly extended Ray's first-hand knowledge of flora and fauna. Back in England, the two friends set to work on their appointed task. In 1670 Ray produced a *Catalogus Plantarum Angliae* ("Catalog of English Plants"). Then, in 1672, Willughby suddenly died, and Ray took up the completion of his portion of their project. In 1676 he published *F. Willughbeii . . . Ornithologia* ("The Ornithology of F. Willughby . . .") under Willughby's name, even though Ray had contributed at least as much as his friend. Ray also completed *F. Willughbeii . . . de Historia Piscium* ("The History of Fish of F. Willughby . . ."; 1685), with the

Royal Society, of which Ray was a fellow, financing its publication.

In 1682 Ray published a *Methodus Plantarum Nova* ("New Method [of Classification] of Plants"; revised in 1703 as the *Methodus Plantarum Emendata . . .*, or "Emended Method of Plants . . ."), his contribution to classification, which insisted on the taxonomic importance of the distinction between monocotyledons and dicotyledons – plants whose seeds germinate with one leaf and with two, respectively. Ray's enduring legacy to botany was the establishment of species as the ultimate unit of taxonomy. On the basis of the *Methodus* he constructed his masterwork, the *Historia Plantarum* ("History of Plants"), three huge volumes that appeared between 1686 and 1704. After the first two volumes, he was urged to compose a complete system of nature. To this end he compiled brief synopses of British and European plants, a *Synopsis Methodica Avium et Piscium* ("Synopsis of Birds and Fish"; published posthumously, 1713), and a *Synopsis Methodica Animalium Quadrupedum et Serpentini Generis* ("Synopsis of Quadrupeds"; 1693). Much of his final decade was spent on a pioneering investigation of insects, published posthumously as *Historia Insectorum* ("History of Insects").

In all this work, Ray contributed to the ordering of taxonomy. Instead of a single feature, he attempted to base his systems of classification on all the structural characteristics, including internal anatomy. By insisting on the importance of the lungs and cardiac structure he effectively established the class of mammals, and he divided insects according to the presence or absence of metamorphoses. Although a truly natural system of taxonomy could not be realized before the age of Darwin (1809–82), Ray's system approached that goal more than the frankly artificial systems of his contemporaries. He was one of the great predecessors who made possible the contributions of Carolus Linnaeus (1707–78) in the following century. While still working on his *Historia Insectorum*, John Ray died at the age of 77.

SIR ISAAC NEWTON (1642–1727)

English physicist and mathematician, who was the culminating figure of the scientific revolution of the seventeenth century.

Born in the hamlet of Woolsthorpe, Lincolnshire, Newton was the only son of a local yeoman, also Isaac Newton, who had died three months before, and of Hannah Ayscough. That same year, at Arcetri near Florence, Galileo Galilei had died; Newton would later pick up his idea of a mathematical science of motion and bring his work to full fruition. A tiny and weak baby, Newton was not expected to survive his first day of life, much less 84 years. Deprived of a father before birth, he soon lost his mother as well, for within two years she married a second time. Her husband, the well-to-do minister Barnabas Smith, left young Isaac with his grandmother and moved to a neighbouring village to raise a son and two daughters.

After his mother was widowed a second time, she determined that her first-born son should manage her now considerable property. It quickly became apparent, however, that this would be a disaster – both for the estate and for Newton – for he could not bring himself to concentrate on rural affairs. Fortunately the mistake was recognized, and Newton was sent back to the grammar school in Grantham to prepare for the university. At the school he apparently gained a firm command of Latin but probably received no more than a smattering of arithmetic. By June 1661, he was ready to matriculate at Trinity College, University of Cambridge, somewhat older than the other undergraduates because of his interrupted education.

When Newton arrived in Cambridge, the movement now known as the scientific revolution was well advanced. Yet the universities of Europe, including Cambridge, continued be the strongholds of outmoded Aristotelianism, which rested on a geocentric view of the universe and dealt with nature in qualitative rather than quantitative terms. Some time during his undergraduate career, Newton discovered the works of the French natural philosopher René Descartes (1596–1650) and the other mechanical philosophers, who, in contrast to Aristotle, viewed physical reality as composed entirely of particles of matter in motion and who held that all the phenomena of nature result from their mechanical interaction. Newton's scientific career had begun.

Although he did not record it, Newton had also begun his mathematical studies. From Descartes' *Geometry* he branched out into the other literature of modern analysis with its application of algebraic techniques to problems of geometry. He then reached back for the support of classical geometry. Within little more than a year he had mastered the literature, and, pursuing his own line of analysis, began to move into new territory. He discovered the binomial theorem, a formula giving the expansion of the powers of sums; and he developed the calculus, a more powerful form of analysis that employs infinitesimal considerations in finding the slopes of curves and areas under curves.

When Newton received the bachelor's degree in April 1665, the most remarkable undergraduate career in the history of university education had passed unrecognized. On his own, without formal guidance, he had sought out the new philosophy and the new mathematics and made them his own – but he had confined the progress of his studies to his notebooks. Then, in 1665, the plague closed the university, and for most of the following two years Newton was forced to stay at his home During the plague years he laid the foundations of the calculus and extended an earlier insight concerning light into an essay, *Of Colours*. It was during this time that he examined the elements of circular motion and, applying his analysis to the moon and the planets, derived the inverse square relation that the radially directed force acting on a planet decreases with the square of its distance from the sun – which was later crucial to the law of universal gravitation. The world heard nothing of these discoveries.

Newton was elected to a fellowship in Trinity College in 1667, after the university reopened. Two years later, Isaac Barrow, Lucasian Professor of Mathematics – who had transmitted Newton's important *De Analysi per Aequationes Numeri Terminorum Infinitas* ("On Analysis by Infinite Series"), written in 1669, to John Collins in London – resigned the chair to devote himself to divinity and recommended Newton to succeed him. The professorship exempted Newton from the necessity of tutoring but imposed the duty of delivering an annual course of lectures. He chose the work he had done in optics as the initial topic; during the following three years (1670–72), his lectures developed the essay *Of Colours* into a form that was later revised to become Book One of his *Opticks* (1704). Through a series of experiments performed in 1665 and 1666, in which the spectrum of a narrow beam was projected onto the wall of a darkened chamber, Newton determined that light is complex and heterogeneous and that the phenomena of colours arise from the analysis, or separation, of the heterogeneous mixture into its simple components. He also concluded that rays refract at distinct angles (hence the prismatic spectrum – a beam of parallel heterogeneous rays analysed by refraction into its component parts by a prism) and that phenomena such as the rainbow are produced by refractive analysis. Because he believed that chromatic aberration could never be eliminated from lenses, Newton turned to reflecting telescopes, and constructed the first one ever built. The heterogeneity of light has been the foundation of physical optics since his time.

There is no evidence that the theory of colours, fully described by Newton in his inaugural lectures at Cambridge, made any impression, just as there is no evidence that aspects of his mathematics and the content of the *Principia*, also pronounced from the podium, made any impression. Rather, the theory of colours, like his later work, was transmitted to the world through the Royal Society, which had been organized in 1660. When Newton was appointed Lucasian professor, his name was probably unknown in the Royal Society; in 1671, however, they heard of his reflecting telescope and asked to see it. Pleased by their enthusiastic reception of the telescope and by his election to the society, Newton volunteered a paper on light and colours early in 1672. On the whole, the paper was also well received, although a few questions and some dissent were heard.

Among the most important dissenters to Newton's paper was Robert Hooke, one of the leaders of the Royal Society who considered himself the master in optics and who wrote a condescending critique of the unknown parvenu. One can understand how the critique would have annoyed a normal man. However, the flaming rage it provoked, with the desire publicly to humiliate Hooke, bespoke the abnormal. Newton was unable rationally to confront criticism. Less than a year after submitting the paper he was so unsettled by the give-and-take of honest discussion that he began to cut his ties, and he withdrew into virtual isolation.

In 1675, during a visit to London, Newton thought he heard Hooke accept his theory of colours. He was emboldened to bring forth a second paper, an examination of the colour phenomena in thin films, which was identical to most of Book Two as it later appeared in the *Opticks*. The purpose of the paper was to explain the colours of solid bodies by showing how light can be analysed into its components by reflection as well as refraction. His explanation of the colours of bodies has not survived, but the paper was significant in demonstrating for the first time the existence of periodic optical phenomena. He discovered the concentric coloured rings in the thin film of air between a lens and a flat sheet of glass; the distance between these concentric rings (now known as Newton's rings) depends on the increasing thickness of the film of air.

A second piece that Newton had sent with the paper of 1675 provoked new controversy. Entitled "An Hypothesis Explaining the Properties of Light", it was in fact a general system of nature. Hooke apparently claimed that Newton had stolen its content from him, and Newton boiled over again. The issue was quickly controlled, however, by an exchange of formal, excessively polite letters that fail to conceal the complete lack of warmth between the men. The rivalry between Newton and Hooke would erupt again later.

In August 1684 Newton was visited by the British astronomer Edmond Halley (1656–1742). Upon learning that Newton had solved the problem of orbital dynamics, he extracted Newton's promise to send the demonstration. Three months later he received a short tract entitled *De Motu* ("On Motion"). Already Newton was at work improving and expanding it. In two and a half years, the tract *De Motu* grew into *Philosophiae Naturalis Principia Mathematica* ("Mathematical Principles of Natural Philosophy"), which is not only

Newton's masterpiece but also the fundamental work for the whole of modern science.

Significantly, *De Motu* did not state the law of universal gravitation and did not contain any of the three Newtonian laws of motion. Only when revising *De Motu* did Newton embrace the principle of inertia (the first law) and arrive at the second law of motion. The second law – the force law – proved to be a precise quantitative statement of the action of the forces between bodies that had become the central members of his system of nature. By quantifying the concept of force, the second law completed the exact quantitative mechanics that has been the paradigm of natural science ever since.

The quantitative mechanics of the *Principia* is not to be confused with its mechanical philosophy. The latter was a philosophy of nature that attempted to explain natural phenomena by means of imagined mechanisms among invisible particles of matter. The mechanics of the *Principia* was an exact quantitative description of the motions of visible bodies. It rested on Newton's three laws of motion: (1) that a body remains in its state of rest unless it is compelled to change that state by a force impressed on it, (2) that the change of motion (the change of velocity times the mass of the body) is proportional to the force impressed, and (3) that to every action there is an equal and opposite reaction.

When the Royal Society received the completed manuscript of Book I of the *Principia* in 1686, Hooke raised the cry of plagiarism based on limited correspondence with Newton in 1679. The charge cannot be sustained in any meaningful sense, but Newton's response to it reveals much about him. Hooke would have been satisfied with a generous acknowledgment; it would have been a graceful gesture to a sick man already well into his decline, and it would have cost Newton nothing.

Newton, instead, went through his manuscript and eliminated nearly every reference to Hooke.

The *Principia* immediately raised Newton to international prominence. In their continuing loyalty to the mechanical ideal, Continental scientists rejected the idea of action at a distance for a generation, but even in their rejection they could not withhold their admiration for the technical expertise revealed by the work.

Almost immediately following the *Principia*'s publication, Newton – a fervent if unorthodox Protestant – helped to lead the resistance of Cambridge to James II's attempt to Catholicize it. As a consequence, he was elected to represent the university in the convention that arranged the revolutionary settlement. In this capacity, he made the acquaintance of a broader group, including the philosopher John Locke (1632–1704) and tasted the excitement of London life. The great bulk of his creative work completed, he sought position at court and in 1696 was appointed warden of the mint.

As warden and then master of the mint, Newton drew a large income, as much as £2,000 per annum. Added to his personal estate, the income left him a rich man at his death. The position, regarded as a sinecure, was treated otherwise by Newton. During the great recoinage there was a need for him to be actively in command, but even afterwards he chose to exercise himself in the office. Above all, he was interested in the problem of counterfeiting. He became the terror of London counterfeiters, sending a goodly number to the gallows and finding in them a socially acceptable target on which to vent the rage that continued to well up within him.

In London, Newton also assumed the role of patriarch of English science. In 1703 he was elected President of the Royal Society. Four years earlier, the French Académie des Sciences ("Academy of Sciences") had named him one of eight foreign associates. He was knighted in 1705, the first occasion on which a scientist was so honoured. During his final years he brought out further editions of his central works. After the first edition of the *Opticks* in 1704, which merely published work done 30 years before, he published a Latin edition in 1706 and a second English edition in 1717–18. In both, the central text was scarcely touched, but he did expand the "Queries" at the end into the final statement of his speculations on the nature of the universe. The second edition of the *Principia*, edited by Roger Cotes in 1713, introduced extensive alterations. A third edition, edited by Henry Pemberton in 1726, added little more. Until nearly the end of his life, Newton presided at the Royal Society and supervised the mint.

GOTTFRIED WILHELM LEIBNIZ (1646–1716)

German philosopher and mathematician, distinguished for his independent invention of the differential and integral calculus.

Leibniz was born into a pious Lutheran family in Leipzig near the end of the Thirty Years' War, which had laid Germany in ruins. As a child he was educated in the Nicolai School but was largely self-taught in the library of his father, who had died in 1652. After completing his legal studies at the University of Leipzig in 1666, Leibniz applied for the degree of doctor of law. He was refused because of his age and consequently left his native city forever. At Altdorf – the university town of the free city of Nürnberg – his dissertation *De Casibus Perplexis* ("On Perplexing Cases") procured him the doctor's degree at once. Liebniz then entered service in the court of the prince elector, the archbishop of Mainz, Johann Philipp von Schönborn, where he was concerned with questions of law and politics.

In 1672 the young jurist was sent on a mission to Paris but was soon left without protectors by the deaths of his German patrons. However, he was now free to pursue his scientific studies. In search of financial support, he constructed a calculating machine and presented it to the Royal Society during his first journey to London, in 1673. Late in 1675 Leibniz laid the foundations of both integral and differential calculus.

In October 1676 he accepted a position in the employment of John Frederick, the Duke of Braunschweig-Lüneburg. John Frederick, a convert to Catholicism from Lutheranism in 1651, had become Duke of Hanover in 1665. He appointed Leibniz librarian, but, beginning in February 1677, Leibniz solicited the post of councillor, which he was finally granted in 1678. It should be noted that he was the only one among the great philosophers of his time who had to earn a living. As a result, he was always a jack-of-all-trades to royalty.

Trying to make himself useful in all ways, Leibniz proposed that education be made more practical and that academies be founded; he worked on hydraulic presses, windmills, lamps, submarines, clocks, and a wide variety of mechanical devices; he devised a means of perfecting carriages and experimented with phosphorus. He also developed a water pump run by windmills, which ameliorated the exploitation of the mines of the Harz Mountains, and he worked in these mines as an engineer frequently from 1680 to 1685.

These many occupations did not stop his work in mathematics: In March 1679 he perfected the binary system of numeration (i.e. using two as a base), and at the end of the same year he proposed the basis for *analysis situs*, now known as general topology, a branch of mathematics that deals with selected properties of collections of related physical or abstract elements. He was also working on his dynamics and his philosophy, which was becoming increasingly anti-Cartesian. In the early 1680s Leibniz continued to perfect his metaphysical system through research into the notion of a universal cause of all being, attempting to arrive at a starting point that would reduce reasoning to an algebra of thought.

In 1685 he was named historian for the House of Brunswick and, on this occasion, *Hofrat* ("Court adviser"). His job was to prove, by means of genealogy, that the princely house had its origins in the House of Este, an Italian princely family, which would allow Hanover to lay claim to a ninth electorate. In search of these documents, Leibniz began travelling in November 1687. Going by way of southern Germany, he arrived in Austria, where he learned that Louis XIV had once again declared a state of war. In Vienna, he was well received by the Emperor; he then went to Italy. Everywhere he went, he met scientists and continued his scholarly work, publishing essays on the movement of celestial bodies and on the duration of things. He returned to Hanover in mid-July 1690.

Until the end of his life, Leibniz continued his duties as historian. He did not, however, restrict himself to a genealogy of the House of Brunswick but enlarged his goal to a history of the Earth, which included such matters as geological events and descriptions of fossils. He searched by way of monuments and linguistics for the origins and migrations of peoples; then for the birth and progress of the sciences, ethics, and politics; and, finally, for the elements of a *historia sacra* ("sacred history"). In this project of a universal history, Leibniz never lost sight of the fact that everything interlocks.

In 1691 Leibniz was named librarian at Wolfenbüttel and propagated his discoveries by means of articles in scientific journals. In 1695 he explained a portion of his dynamic theory of motion in the *Système nouveau* ("New System"), which treated the relationship of substances and the pre-established harmony between the soul and the body. He asserted that God does not need to bring about man's action by means of his thoughts, as Malebranche maintained, or to wind some sort of watch in order to reconcile the two; rather, the Supreme Watchmaker has so exactly matched body and soul that they correspond – they give meaning to each other – from the beginning. In 1697, *De Rerum Originatione* ("On the Ultimate Origin of Things") tried to prove that the ultimate origin of things can be none other than God. In 1698, *De Ipsa Natura* ("On Nature Itself") explained the internal activity of nature in terms of Leibniz's theory of dynamics.

All of these writings opposed Cartesianism, which was judged to be damaging to faith. Plans for the creation of German academies followed in rapid succession. With the help of the electress Sophia Charlotte, daughter of Ernest Augustus and soon to become the first queen of Prussia (January 1701), the German Academy of Sciences in Berlin was founded on 11 July 1700.

LEONHARD EULER (1707–1783)

Swiss mathematician and physicist,
one of the founders of pure mathematics.

Euler was born in Basel, Switzerland, and studied at the University of Basel, where he received a master's degree. His mathematical ability earned him the esteem of Johann Bernoulli, one of the first mathematicians in Europe at that time, and of his sons Daniel and Nicolas. In 1727 he moved to St Petersburg, where he became an associate of the St Petersburg Academy of Sciences and in 1733 succeeded Daniel Bernoulli to the chair of mathematics.

By means of his numerous books and memoirs that he submitted to the academy, Euler carried integral calculus to a higher degree of perfection, developed the theory of trigonometric and logarithmic functions, reduced analytical operations to a greater simplicity, and threw new light on nearly all parts of pure mathematics. Overtaxing himself, in 1735 Euler lost the sight of one eye. Invited by Frederick the Great in 1741, he became a member of the Berlin Academy, where for 25 years he produced a steady stream of publications – many of which he contributed to the St Petersburg Academy, which granted him a pension.

Euler's textbooks in calculus, *Institutiones calculi differentialis* ("Institutions of Differential Calculus"; 1755 and *Institutiones calculi integralis* ("Institutions of Integral Calculus"; 1768–70, have served as prototypes to the present because they contain formulae of differentiation and numerous methods of indefinite integration, many of which Euler invented himself, for determining the work done by a force and for solving geometric problems. Euler also made advances in the theory of linear differential equations, which are useful in solving problems in physics. Thus, he enriched mathematics with substantial new concepts and techniques. He introduced many current notations, such as Σ for the sum; the symbol e for the base of natural logarithms; a, b, and c for the sides of a triangle and A, B, and C for the opposite angles; the letter "f" with parentheses for a function; the use of the symbol π (pi) for the ratio of circumference to diameter in a circle; and i for $\sqrt{-1}$. His interests were broad, and his *Lettres à une princesse d'Allemagne* ("Letters to a German Princess") in 1768–72 were an admirably clear exposition of the basic principles of mechanics, optics, acoustics, and physical astronomy.

After Frederick the Great became less cordial toward him, Euler in 1766 accepted the invitation of Catherine II to return to Russia. Soon after his arrival in St Petersburg, a cataract formed in his remaining good eye, and he spent the last years of his life in total blindness. Despite this tragedy, his productivity continued undiminished, sustained by an uncommon memory and a remarkable facility in mental computations.

CAROLUS LINNAEUS (1707–1778)

Swedish naturalist and explorer who created a uniform system (binomial nomenclature) for naming natural genera and species of organisms.

Linnaeus was the son of a curate and grew up in Småland, a poor region in southern Sweden. In 1727 he began his studies in medicine at Lund University, but transferred to Uppsala University in 1728. Because of his financial situation, he could visit only a few lectures; however, the university professor Olof Celsius provided Linnaeus access to his library. From 1730 to 1732 he was able to subsidize himself by teaching botany in the university garden of Uppsala.

In 1732 the Uppsala Academy of Sciences sent Linnaeus on a research expedition to Lapland. After his return in the autumn of that year, he gave private lectures in botany and mineral assaying. At the time, it was necessary for Swedish medical students to complete their doctoral degrees abroad in order to open a successful medical practice in their homeland, and so Linnaeus found a patron who would pay for his studies in the Netherlands.

In May 1735, he completed his examinations and received his medical degree. He then journeyed to Leiden, where he sought patronage for the publication of his numerous manuscripts. He was immediately successful, and his *Systema Naturae* ("The System of Nature") was published only a few months later with financial support from Jan Frederik Gronovius, senator of Leiden, and Isaac Lawson, a Scottish physician.

This folio volume of only 11 pages presented a hierarchical classification, or taxonomy, of the three kingdoms of nature: stones, plants, and animals. Each kingdom was subdivided into classes, orders, genera, species, and varieties. This hierarchy of taxonomic ranks replaced traditional systems of biological classification that were based on mutually exclusive divisions, or dichotomies. Linnaeus's classification system has survived in biology, although additional ranks, such as families, have been added to accommodate growing numbers of species.

In particular, it was the botanical section of *Systema Naturae* that built Linnaeus's scientific reputation. After reading essays on sexual reproduction in plants by the French botanist Sébastian Vaillant and the German botanist Rudolph Jacob Camerarius, Linnaeus had become convinced of the idea that all organisms reproduce sexually. As a result, he expected each plant to possess male and female sexual organs (stamens and pistils), or "husbands and wives", as he also put it. On this basis, he designed a simple system of distinctive characteristics to classify each plant. The number and position of the stamens, or husbands, determined the class to which it belonged, whereas the number and position of pistils, or wives, determined the order.

Linnaeus did not consider the sexual system to be his main contribution toward the "reformation of botany" to which he aspired. Rather, this came in the form of a booklet, the *Fundamenta Botanica* ("The Foundations of Botany"; 1736), that framed the principles and rules to be followed in the classification and naming of plants.

In 1735 Linnaeus met Hermann Boerhaave, who introduced him to George Clifford, a local English merchant and banker who had close connections to the Dutch East India Company. Impressed by Linnaeus's knowledge, Clifford offered Linnaeus a position as curator of his botanical garden. Linnaeus accepted the position and used this opportunity to expand certain chapters of the *Fundamenta Botanica* in separate publications: the *Bibliotheca Botanica* ("The Library of Botany"; 1736); *Critica Botanica* ("A Critique of Botany"; 1737), on botanical nomenclature; and *Classes Plantarum* ("Classes of Plants"; 1738). He applied the theoretical framework laid down in these books in two further publications: *Hortus Cliffortianus* ("Clifford's Garden"; 1737), a catalogue of the species contained in Clifford's collection; and the *Genera Plantarum* ("Genera of Plants"; 1737), which modified and updated definitions of plant genera first offered by Joseph Pitton de Tournefort (1656–1708). *Genera Plantarum* was considered by Linnaeus to be his crowning taxonomic achievement.

In contrast to earlier attempts by other botanists at generic definition, which proceeded by a set of arbitrary divisions (as did his own sexual system), *Genera Plantarum* presented a system based on what Linnaeus called the "natural characters" of genera – morphological descriptions of all the parts of flower and fruit. A system based on natural characters could accommodate the growing number of new species – often possessing different morphological features – pouring into Europe from its overseas trading posts and colonies.

Linnaeus returned to Sweden in 1738 and began a medical practice in Stockholm. He practised medicine until the early

1740s but longed to return to his botanical studies. A position became available at Uppsala University, and he received the chair in medicine and botany there in 1742. Linnaeus built his further career upon the foundations he laid in the Netherlands, using his international contacts to create a network of correspondents that provided him with seeds and specimens from all over the world. He then incorporated this material into the botanical garden at Uppsala, and these acquisitions helped him develop and refine the empirical basis for revised and enlarged editions of his major taxonomic works.

Linnaeus's most lasting achievement was the creation of binomial nomenclature, the system of formally classifying and naming organisms according to their genus and species. In contrast to earlier names that were made up of diagnostic phrases, binomial names (or "trivial" names as Linnaeus himself called them) conferred no prejudicial information about the species named. Rather, they served as labels by which a species could be universally addressed. This naming system was also implicitly hierarchical, as each species is classified within a genus. The first use of binomial nomenclature by Linnaeus occurred in the context of a small project in which students were asked to identify the plants consumed by different kinds of cattle. In this project, binomial names served as a type of shorthand for field observations. Despite the advantages of this naming system, binomial names were used consistently in print by Linnaeus only after the publication of the *Species Plantarum* ("Species of Plants"; 1753).

In his own lifetime Linnaeus became something of an institution in himself, as naturalists everywhere had to address him directly or at least his work in order to determine whether specimens in their collections were indeed new species. The rules of nomenclature that he put forward in his *Philosophia Botanica* ("Philosophy of Botany"; 1751) rested on a recognition of the "law of priority" – the rule stating that the first properly published name of a species or genus takes precedence over all other proposed names. These rules became firmly established in the field of natural history and also formed the backbone of international codes of nomenclature, such as the Strickland Code (1842), created for the fields of botany and zoology in the mid-nineteenth century. The first edition of the *Species Plantarum* and the 10th edition of the *Systema Naturae* (1758) are the agreed starting points for botanical and zoological nomenclature, respectively.

GEORGES-LOUIS LECLERC, COMTE DE BUFFON (1707–1788)

French naturalist, remembered for his comprehensive work on natural history, Histoire naturelle, générale et particulière.

Leclerc was born in Montbard, France. At the College of Godrans in Dijon, which was run by the Jesuits, Leclerc seems to have been only an average student, but one with a marked taste for mathematics. His father wanted him to have a legal career, and in 1723 he began the study of law. In 1728, however, he went to Angers, where he seems to have studied medicine and botany as well as mathematics.

He was forced to leave Angers after a duel and took refuge at Nantes, where he lived with a young Englishman, the Duke of Kingston. The two young men travelled to Italy, arriving in Rome at the beginning of 1732. They also visited England, and while there Buffon was elected a member of the Royal Society.

The death of his mother called him back to France. He settled down on the family estate at Montbard, where he undertook his first research in the calculus of probability and in the physical sciences. Buffon at that time was particularly interested in questions of plant physiology, and in 1735 he published a translation of Stephen Hales's *Vegetable Staticks*, in the preface of which he developed his conception of scientific method. He also made researches on the properties of timbers and their improvement in his forests in Burgundy.

In 1739, at the age of 32, he was appointed keeper of the Jardin du Roi (the royal botanical garden, now the Jardin des Plantes) and of the museum that formed part of it, where he was charged to undertake a catalogue of the royal collections in natural history. The ambitious Buffon transformed the task into an account of the whole of nature. This became his great work, *Histoire naturelle, générale et particulière* ("Natural History, General and Particular"; 1749–1804), which was the first modern attempt systematically to present all existing knowledge in the fields of natural history, geology, and anthropology in a single publication.

Buffon's *Histoire naturelle* was translated into various languages and widely read throughout Europe. The first edition is still highly prized by collectors for the beauty of its illustrations. Although Buffon laboured arduously on it – he spent eight months of the year on his estate at Montbard, working up to 12 hours a day – he was able to publish only 36 of the proposed 50 volumes before his death. In the preparation of the first 15 volumes, which appeared in 1749–67, he was assisted by Louis J.M. Daubenton and several other associates.

The next seven volumes formed a supplement to the preceding ones and appeared in 1774–89; the most famous section, *Époques de la nature* ("Epochs of Nature"; 1778), being contained in the fifth. They were succeeded by nine volumes on birds (1770–83), then by five volumes on minerals (1783–8). The remaining eight volumes, which complete the first edition, were done by the Comte de Lacépède after Buffon's death: they covered reptiles, fishes, and cetaceans. To keep the descriptions of the animals from becoming monotonous, Buffon interspersed them with philosophic discussions on nature, the degeneration of animals, the nature of birds, and other topics.

Buffon was elected to the French Academy, where, on 25 August 1753, he delivered his celebrated *Discours sur le style* ("Discourse on Style"), containing the line *"Le style c'est l'homme même"* ("Style is the man himself"). He was also treasurer to the Academy of Sciences and, in 1773, was created a count. During the brief trips he made each year to Paris he frequented the literary and philosophical salons. He enjoyed his life at Montbard, living in contact with nature and the peasants and managing his properties himself. He built a menagerie and a large aviary there and transformed one of his outbuildings into a laboratory.

In some areas of natural science Buffon had a lasting influence. He was the first to reconstruct geological history in a series of stages, in *Époques de la nature* . With his notion of lost species he opened the way to the development of paleontology. He was the first to propose the theory that the planets had been created in a collision between the sun and a comet. While his great project opened up vast areas of knowledge that were beyond his powers to encompass, his *Histoire naturelle* was the first work to present the previously isolated and apparently disconnected facts of natural history in a generally intelligible form.

JEAN LE ROND
D'ALEMBERT (1717–1783)

French mathematician and scientist who gained
a considerable reputation as a contributor to
and editor of the famous *Encyclopédie*.

The illegitimate son of a famous hostess, Mme de Tencin, and
one of her lovers, the chevalier Destouches-Canon, d'Alembert
was abandoned on the steps of the Parisian church of Saint-
Jean-le-Rond, from which he derived his Christian name. He
spent two years studying law and became an advocate in 1738,
although he never practised. After taking up medicine for a
year, he finally devoted himself to mathematics – "the only
occupation", he said later, "which really interested me." Apart
from some private lessons, d'Alembert was almost entirely self-
taught.

In 1739 he read his first paper to the Academy of Sciences, of
which he became a member in 1741. In 1743, at the age of 26,
he published his important *Traité de dynamique* ("Treatise on
Dynamics"), a fundamental treatise on dynamics containing
the famous "d'Alembert's principle", which states that Newton's third law of motion (i.e., for every action there is an equal
and opposite reaction) is true for bodies that are free to move
as well as for bodies rigidly fixed. Other mathematical works
followed very rapidly: in 1744 he applied his principle to the
theory of equilibrium and motion of fluids, in his *Traité de
l'équilibre et du mouvement des fluides*. This discovery was
followed by the development of partial differential equations,
a branch of the theory of calculus, the first papers on which
were published in his *Réflexions sur la cause générale des vents*
("Reflections on the General Cause of Winds", 1747). It won
him a prize at the Berlin Academy, to which he was elected the

same year. In 1747 he applied his new calculus to the problem of vibrating strings, in his *Recherches sur les cordes vibrantes*; in 1749 he furnished a method of applying his principles to the motion of any body of a given shape; and in 1749 he found an explanation of the precession of the equinoxes (a gradual change in the orientation of the Earth's axis), determined its characteristics, and explained the phenomenon of the nutation (nodding) of the Earth's axis, in *Recherches sur la précession des équinoxes et sur la nutation de l'axe de la terre.*

Meanwhile, d'Alembert began an active social life and frequented well-known salons, where he acquired a considerable reputation as a witty conversationalist and mimic. Like his fellow Philosophes – those thinkers, writers, and scientists who believed in the sovereignty of reason and nature (as opposed to authority and revelation) and rebelled against old dogmas and institutions – he turned to the improvement of society. A rationalist thinker in the free-thinking tradition, he opposed religion and stood for tolerance and free discussion; in politics the Philosophes sought a liberal monarchy with an "enlightened" king who would supplant the old aristocracy with a new, intellectual aristocracy.

Science, the only real source of knowledge, had to be popularized for the benefit of the people, and it was in this tradition that d'Alembert became associated with the *Encyclopédie* in about 1746. When the original idea of a translation into French of Ephraim Chambers's English *Cyclopædia* was replaced by that of a new work under the general editorship of the Philosophe Denis Diderot, d'Alembert was made editor of the mathematical and scientific articles. In fact, he not only helped with the general editorship and contributed articles on other subjects but also tried to secure support for the enterprise in influential circles. He wrote the *Discours préliminaire* ("Preliminary

Discourse") that introduced the first volume of the work in 1751. This was a remarkable attempt to present a unified view of contemporary knowledge, tracing the development and interrelationship of its various branches and showing how they formed coherent parts of a single structure. The second section of the *Discours* was devoted to the intellectual history of Europe from the time of the Renaissance.

In 1752 d'Alembert wrote a preface to Volume III, which was a vigorous rejoinder to the *Encyclopédie*'s critics, while an *Éloge de Montesquieu* ("Elegy to Montesquieu"), which served as the preface to Volume V (1755), skilfully but somewhat disingenuously presented Montesquieu as one of the *Encyclopédie*'s supporters. Montesquieu had, in fact, refused an invitation to write the articles "Democracy" and "Despotism", and the promised article on "Taste" remained unfinished at his death in 1755.

In 1756 d'Alembert went to stay with Voltaire at Geneva, where he also collected information for an *Encyclopédie* article, "Genève", which praised the doctrines and practices of the Genevan pastors. When it appeared in 1757 it aroused angry protests in Geneva because it affirmed that many of the ministers no longer believed in Christ's divinity, and also advocated (probably at Voltaire's instigation) the establishment of a theatre. This article prompted Rousseau, who had contributed the articles on music to the *Encyclopédie*, to argue in his *Lettre à d'Alembert sur les spectacles* ("Letter to d'Alembert on the Theatre"; 1758) that the theatre is invariably a corrupting influence. D'Alembert himself replied with an incisive but not unfriendly *Lettre à J.-J. Rousseau, citoyen de Genève* ("Letter to J.-J. Rousseau, Citizen of Geneva"). Gradually discouraged by the growing difficulties of the enterprise, d'Alembert gave up his share of the editorship at the beginning of 1758, thereafter limiting his commitment to the production of mathematical and scientific articles.

D'Alembert was elected to the French Academy in 1754 and proved himself to be a zealous member, working hard to enhance the dignity of the institution in the eyes of the public and striving steadfastly for the election of members sympathetic to the cause of the Philosophes. His personal position became even more influential in 1772 when he was made permanent secretary.

HENRY CAVENDISH (1731–1810)

The greatest English experimental and
theoretical chemist and physicist of his age.

Cavendish was born in Nice, France. His father, Lord Charles Cavendish, was the third son of the Duke of Devonshire, and his mother (née Ann Grey) was the fourth daughter of the Duke of Kent; they were resident in France at the time of Henry's birth. Henry had no title, but is often referred to as "the Honourable Henry Cavendish". He went to the Hackney Academy, a private school near London, and in 1748 entered Peterhouse College, University of Cambridge, where he remained for three years before leaving without taking a degree. He then lived with his father in London.

His father lived a life of service, first in politics and then increasingly in science, especially in the Royal Society. In 1758 he took Henry to meetings of the Royal Society and also to dinners of the Royal Society Club. In 1760 Henry Cavendish was elected to both these groups, and was assiduous in his attendance. He was active in the Council of the Royal Society, to which he was elected in 1765, and his interest and expertise in the use of scientific instruments led him to head a committee to review the Royal Society's meteorological instruments and to help assess the instruments of the Royal Greenwich Observatory. Other committees on which he served included the committee of papers, which chose the papers for publication in the *Philosophical Transactions*; and the committees for the transit of Venus, for the gravitational attraction of mountains, and for the scientific instructions for Constantine Phipps's expedition in search of the North Pole and the Northwest Passage. In 1773 Henry joined his father as an elected trustee of the British Museum, to which he devoted a good deal of time and effort. Cavendish became a manager of the Royal Institution of Great Britain in 1800, soon after its establishment, and took an active interest, especially in the laboratory, where he observed and helped in Humphry Davy's chemical experiments.

Cavendish had a laboratory in the house where he lived with his father, and here he conducted his first electrical and chemical experiments. In 1783 his father died, leaving almost all of his very substantial estate to Henry. Henry bought another house in town and also a house in Clapham Common, south of London. The London house contained the bulk of his library, while most of his experiments were carried out at Clapham Common.

Cavendish was a shy man who was uncomfortable in society and avoided it when he could. He conversed little, always dressed in an old-fashioned suit, and developed no known deep personal attachments outside his family. At about the time of his father's death, Cavendish began to work closely with the physician and scientist Charles Blagden, an association that helped Blagden enter fully into London's scientific society. In return, Blagden helped to keep the world at a distance from Cavendish. Cavendish published no books and few papers, but he achieved much. Several areas of research, including mechanics, optics, and magnetism, feature extensively in his manuscripts, but scarcely feature in his published work.

His first publication, in 1766, was a combination of three short chemistry papers on "factitious airs", or gases produced in the laboratory. He produced "inflammable air" (hydrogen) by dissolving metals in acids, and "fixed air" (carbon dioxide) by dissolving alkalis in acids, and he collected these and other gases in bottles inverted over water or mercury. He then measured their solubility in water and their specific gravity and noted their combustibility. Cavendish was awarded the Royal Society's Copley Medal for this paper. Gas chemistry was of increasing importance in the latter half of the eighteenth century and became crucial for the reform of chemistry, generally known as the chemical revolution, brought about by the Frenchman Antoine-Laurent Lavoisier (1743–94).

In 1783 Cavendish published a paper on eudiometry (the measurement of the quality of gases for breathing). He described a new eudiometer of his own invention, with which he achieved the best results to date, using what in other hands had been the inexact method of measuring gases by weighing them. He next published a paper on the production of water by burning inflammable air (hydrogen) in "dephlogisticated air" (now known to be oxygen), the latter a constituent of atmospheric air. Cavendish concluded that dephlogisticated air was dephlogisticated water and that hydrogen was either pure "phlogiston" or phlogisticated water. He reported these findings to Joseph Priestley, an English clergyman and scientist, no later than March 1783, but did not publish them until the following year. In 1783 the Scottish inventor James Watt published a paper on the composition of water. Cavendish had performed the experiments first but published second, and controversy about priority ensued.

In 1785 Cavendish carried out an investigation of the composition of common (i.e. atmospheric) air – obtaining, as usual, impressively accurate results. He observed that, when he had determined the amounts of phlogisticated air (nitrogen) and dephlogisticated air (oxygen), there remained a volume of gas amounting to one hundred-and-twentieth of the original volume of common air. A hundred years later, in the 1890s, two British physicists, William Ramsay and Lord Rayleigh, realized that their newly discovered inert gas, argon, was responsible for Cavendish's problematic residue, and that he had not made an error. What he had done was perform rigorous quantitative experiments, using standardized instruments and methods, aimed at reproducible results; taken the mean of the result of several experiments; and identified and allowed for sources of error.

Cavendish, as indicated, used the language of the old phlogiston theory in chemistry. (This theory held that combustible substances released phlogiston, a "fiery substance", when burned.) In 1787 he became one of the earliest scientists outside France to convert to the new oxygen theory of combustion of Lavoisier, although he remained sceptical about the nomenclature of the new theory. He also objected to Lavoisier's identification of heat as having a material or elementary basis. Working within the framework of Newtonian mechanics, Cavendish had tackled the problem of the nature of heat in the 1760s, explaining heat as the result of the motion of matter. In 1783 he published a paper on the temperature at which mercury freezes and in that paper made use of the idea of latent heat – though he did not use the term because he believed that it implied acceptance of a material theory of heat. He made his objections explicit in his 1784 paper on air. He went on to develop a general theory of heat, and the manuscript of that theory has been persuasively dated to the late 1780s. His theory was at once mathematical and mechanical: it contained the principle of the conservation of heat (later understood as an instance of conservation of energy) and even contained the concept (although not the label) of the mechanical equivalent of heat.

Cavendish also worked out a comprehensive theory of electricity. Like his theory of heat, this was mathematical in form and based on precise quantitative experiments. In 1771 he published an early version of his theory, based on an expansive electrical fluid that exerted pressure. He demonstrated that if the intensity of electric force was inversely proportional to distance, then the electric fluid in excess of that needed for electrical neutrality would lie on the outer surface of an electrified sphere; and he confirmed this experimentally. Cavendish continued to work on electricity after this initial paper, but he published no more on the subject.

The most famous of Cavendish's experiments, published in 1798, was to determine the density of the Earth. His apparatus for weighing the world was a modification of the Englishman John Michell's torsion balance. The balance had two small lead balls suspended from the arm of a torsion balance and two much larger stationary lead balls. Cavendish calculated the attraction between the balls from the period of oscillation of the torsion balance, and then used this value to calculate the density of the Earth. What was extraordinary about this experiment was its elimination of every source of error and every factor that could disturb the experiment and its precision in measuring an astonishingly small attraction – a mere 1/50,000,000 of the weight of the lead balls. The result that Cavendish obtained for the density of the Earth is within 1 per cent of the currently accepted figure. The combination of painstaking care, precise experimentation, thoughtfully modified apparatus, and fundamental theory carries Cavendish's unmistakable signature. It is fitting that the University of Cambridge's great physics laboratory is named the Cavendish Laboratory.

Cavendish's electrical papers from the *Philosophical Transactions* of the Royal Society have been reprinted, together with most of his electrical manuscripts, in *The Scientific Papers of the Honourable Henry Cavendish, F.R.S.* (1921). Cavendish remained active in science and healthy in body almost until the end of his life.

JOSEPH PRIESTLEY (1733–1804)

English clergyman, political theorist, and physical scientist, best remembered for his contribution to the chemistry of gases.

Priestley was born into a family of moderately successful wool-cloth makers in the Calvinist stronghold of West Riding, Yorkshire. He entered the Dissenting Academy at Daventry, Northamptonshire, in 1752. Dissenters, so named for their unwillingness to conform to the Church of England, were prevented by the Act of Uniformity (1662) from entering English universities. Between 1755 and 1761 Priestley ministered at Needham Market, Suffolk, and at Nantwich, Cheshire. In 1761 he became tutor in languages and literature at the Warrington Academy, Lancashire. He was ordained a Dissenting minister in 1762.

Priestley's interest in science intensified in 1765, when he met the American scientist and statesman Benjamin Franklin

(1706–90), who encouraged him to publish *The History and Present State of Electricity, with Original Experiments* (1767). In this work, Priestley used history to show that scientific progress depended more on the accumulation of "new facts" that anyone could discover than on the theoretical insights of a few men of genius. This view of scientific methodology shaped Priestley's electrical experiments, in which he anticipated the inverse square law of electrical attraction, discovered that charcoal conducts electricity, and noted the relationship between electricity and chemical change. On the basis of these experiments, in 1766 he was elected a member of the Royal Society. This line of investigation inspired him to develop "a larger field of original experiments" in areas other than electricity.

Upon his return to the ministry at Mill Hill Chapel, Leeds, in 1767, Priestley began intensive experimental investigations into chemistry. Between 1772 and 1790 he published six volumes of *Experiments and Observations on Different Kinds of Air* and more than a dozen articles in the Royal Society's *Philosophical Transactions* describing his experiments on gases, or "airs", as they were then called. British pneumatic chemists had previously identified three types of gases: air, carbon dioxide (fixed air), and hydrogen (inflammable air). Priestley incorporated an explanation of the chemistry of these gases into the phlogiston theory, proposed by George Ernst Stahl (1660–1734), which held that combustible substances released a "fiery substance" called phlogiston during burning.

Priestley discovered ten new gases: nitric oxide (nitrous air), nitrogen dioxide (red nitrous vapour), nitrous oxide (inflammable nitrous air, later called "laughing gas"), hydrogen chloride (marine acid air), ammonia (alkaline air), sulphur dioxide (vitriolic acid air), silicon tetrafluoride (fluor acid air), nitrogen (phlogisticated air), oxygen (dephlogisticated air, independently co-discovered by Carl Wilhelm Scheele), and a gas later identified as carbon monoxide. Priestley's experimental success resulted predominantly from his ability to design ingenious apparatuses and his skill in their manipulation. For his work on gases, he was awarded the Royal Society's prestigious Copley Medal in 1773.

Priestley viewed his scientific pursuits as consistent with the commercial and entrepreneurial interests of English Dissenters. He embraced the argument of the seventeenth-century statesman and natural philosopher Francis Bacon (1561–1626) that social progress required the development of a science-based commerce. This view was reinforced when he moved to become a preacher at the New Meeting House in Birmingham in 1780, and became a member of the Lunar Society, an elite group of local gentlemen, Dissenters, and industrialists (including Josiah Wedgwood, Erasmus Darwin, James Watt, and Matthew Boulton), who applied the principles of science and technology to the solving of problems experienced in eighteenth-century urban life.

Priestley's lasting reputation in science is founded upon the discovery he made on 1 August 1774, when he obtained a colourless gas by heating red mercuric oxide. Finding that a candle would burn and that a mouse would thrive in this gas, he called it "dephlogisticated air", based upon the belief that ordinary air became saturated with "phlogiston" once it could no longer support combustion and life. Priestley was not yet sure, however, that he had discovered a "new species of air".

The following October he accompanied his patron, Shelburne, on a journey through Belgium, Holland, Germany, and France, where in Paris he informed the French chemist Antoine-Laurent Lavoisier how he obtained the new "air". This meeting between the two scientists was highly significant for the future of chemistry. Lavoisier immediately repeated

Priestley's experiments and, between 1775 and 1780, conducted intensive investigations from which he derived the elementary nature of oxygen, recognized it as the "active" principle in the atmosphere, interpreted its role in combustion and respiration, and gave it its name. Lavoisier's pronouncements of the activity of oxygen revolutionized chemistry.

JAMES WATT (1736–1819)

Scottish instrument maker and inventor
whose steam engine contributed substantially
to the Industrial Revolution.

Watt was born in Greenock, Renfrewshire, Scotland, where his father ran a successful ship- and house-building business. His father's workshops were the source of a part of James's education: here, with his own tools, bench, and forge, he made models (e.g. of cranes and barrel organs) and grew familiar with ships' instruments.

Deciding at the age of 17 to be a mathematical-instrument maker, Watt first went to Glasgow, where one of his mother's relatives taught at the university, then, in 1755, to London, where he found a master to train him. Although his health broke down within a year, he had learned enough in that time "to work as well as most journeymen." Returning to Glasgow, he opened a shop in 1757 at the university and made mathematical instruments (e.g. quadrants, compasses, and scales).

While repairing a model Newcomen steam engine in 1764, Watt was dissatisfied by its waste of steam. In May 1765, after wrestling with the problem of improving it, he suddenly came upon a solution – the separate condenser, his first and greatest invention. Watt had realized that the loss of latent heat (the heat involved in changing the state of a substance, e.g. solid or liquid) was the worst defect of the Newcomen engine and that therefore condensation must be effected in a chamber distinct from the cylinder but connected to it. Shortly afterwards he met John Roebuck, the founder of the Carron Works, who urged him to make an engine. He entered into partnership with Roebuck in 1768, after having made a small test engine with the help of loans from a friend, Joseph Black. The following year Watt took out the famous patent for "A New Invented Method of Lessening the Consumption of Steam and Fuel in Fire Engines."

Meanwhile, Watt in 1766 became a land surveyor, and for the next eight years he was continuously busy marking out routes for canals in Scotland – work that prevented his making further progress with the steam engine. After Roebuck went bankrupt in 1772, Matthew Boulton, the manufacturer of the Soho Works in Birmingham, took over a share in Watt's patent. Bored with surveying and with Scotland, Watt moved to Birmingham in 1774. After Watt's patent was extended by an Act of Parliament, he and Boulton in 1775 began a partnership that lasted 25 years. Boulton's financial support made possible rapid progress with the engine. In 1776 two engines were installed, one for pumping water in a Staffordshire colliery; the other for blowing air into the furnaces of John Wilkinson, the famous ironmaster.

During the next five years, until 1781, Watt spent long periods in Cornwall, where he installed and supervised numerous pumping engines for the copper and tin mines, the managers of which wanted to reduce fuel costs. By 1780 he was doing well financially, though Boulton still had problems raising capital. In the following year Boulton, foreseeing a new market in the corn, malt, and cotton mills, urged Watt to invent a rotary motion for the steam engine, to replace the reciprocating action of the original. He did this in 1781 with his so-called "sun-and-planet" gear, by means of which a shaft produced two revolutions for each cycle of the engine.

In 1782, at the height of his inventive powers, Watt patented the double-acting engine, in which the piston pushed as well as pulled. The engine required a new method of rigidly connecting the piston to the beam. He solved this problem in 1784 with his invention of the parallel motion – an arrangement of connected rods that guided the piston rod in a perpendicular motion – which he described as "one of the most ingenious, simple pieces of mechanism I have contrived." Four years later came his application of the centrifugal governor for automatic control of the speed of the engine, at Boulton's suggestion, and in 1790 his invention of a pressure gauge virtually completed the Watt engine. Demands for his engine came quickly from paper, flour, cotton, and iron mills; distilleries; and canals and waterworks. By 1790 Watt was a wealthy man, having received £76,000 in royalties on his patents in 11 years.

The steam engine did not absorb all his attention, however. He was a member of the Lunar Society in Birmingham, a group of writers and scientists who wished to advance the sciences and the arts. Watt experimented on the strength of materials, and he was often involved in legal proceedings to protect his patents. In 1785 he and Boulton were elected fellows of the Royal Society. Watt then began to take holidays, bought an estate at Doldowlod, Radnorshire, and from 1795 gradually withdrew from business. His achievements were amply recognized in his lifetime: he was made doctor of laws of the University of Glasgow in 1806 and a foreign associate of the French Academy of Sciences in 1814. He was also offered a baronetcy, which he declined.

LUIGI GALVANI (1737–1798) AND CONTE ALESSANDRO VOLTA (1745–1827)

Italian physicists who made pioneering
studies of electrical phenomena in
animal tissue and between metals.

Galvani was born in Bologna. He followed his father's pre-
ference for medicine by attending the University of Bologna,
graduating in 1759. On obtaining the Doctor of Medicine
degree, with a thesis (1762), *De ossibus*, on the formation and
development of bones, he was appointed lecturer in anatomy
at the university and professor of obstetrics at the separate
Institute of Arts and Sciences. Beginning with his doctoral
thesis, his early research was in comparative anatomy – as
indicated by his lectures on the anatomy of the frog in 1773
and on electrophysiology in the late 1770s. Following the
acquisition of an electrostatic machine (a large device for
making sparks) and a Leyden jar (a device used to store static
electricity), he began to experiment with muscular stimulation
by electrical means. His notebooks indicate that, from the
early 1780s, animal electricity remained his major field of
investigation.

Numerous ingenious observations and experiments have
been credited to Galvini: in 1786, for example, he obtained
muscular contraction in a frog by touching its nerves with a
pair of scissors during an electrical storm. After a visitor to his
laboratory caused the legs of a skinned frog to kick when a
scalpel touched a lumbar nerve of the animal while an electrical machine was activated, Galvani assured himself by
further experiments that the twitching was, in fact, related
to the electrical action.

He delayed the announcement of his findings until 1791, when he published his essay *De Viribus Electricitatis in Motu Musculari Commentarius* ("Commentary on the Effect of Electricity on Muscular Motion"). He concluded that animal tissue contained a heretofore neglected innate, vital force – which he termed "animal electricity", believing it was a form of electricity – which activated nerve and muscle when spanned by metal probes. He considered the brain to be the most important organ for the secretion of this "electric fluid" and the nerves to be conductors of the fluid to the nerve and muscle, the tissues of which act as did the outer and inner surfaces of the Leyden jar.

Galvani's scientific colleagues generally accepted his views. Alessandro Volta, a physicist from Como, Lombardy, who was professor of physics at the University of Pavia, however, was not convinced by the analogy between the muscle and the Leyden jar. Deciding that the frog's legs served only as an indicating electroscope, Volta held that the contact of dissimilar metals was the true source of stimulation: he referred to the electricity so generated as "metallic electricity" and decided that the muscle, by contracting when touched by metal, resembled the action of an electroscope. Furthermore, Volta said that if two dissimilar metals in contact both touched a muscle, agitation would also occur and increase with the dissimilarity of the metals. Thus Volta rejected the idea of an "animal electric fluid", replying that the frog's legs responded to differences in metal temper, composition, and bulk. Galvani refuted this by obtaining muscular action with two pieces of the same material. The ensuing controversy was without personal animosity, however; Galvani's gentle nature and Volta's high principles precluded any harshness between them. Volta, who coined the term "galvanism", said of Galvani's work that "it contains one of the most beautiful and most surprising discoveries." Nevertheless, partisan groups rallied to both sides.

In retrospect, Galvani and Volta are both seen to have been partly right and partly wrong. Galvani was correct in attributing muscular contractions to an electrical stimulus, but wrong in identifying it as an "animal electricity". Volta correctly denied the existence of an "animal electricity" but was wrong in implying that every electrophysiological effect requires two different metals as sources of current. Galvani, shrinking from the controversy over his discovery, continued his work as teacher, obstetrician, and surgeon, treating both wealthy and needy without regard to fee. In 1794 he offered a defence of his position in an anonymous book, *Dell'uso e dell'attività dell'arco conduttore nella contrazione dei muscoli* ("On the Use and Activity of the Conductive Arch in the Contraction of Muscles"), the supplement of which described muscular contraction without the need of any metal. He caused a muscle to contract by touching the exposed muscle of one frog with a nerve of another, and thus established for the first time that bioelectric forces exist within living tissue.

Galvani's work opened the way to new research in the physiology of muscle and nerve and to the entire subject of electrophysiology. It also provided the major stimulus for Volta's studies of electricity and Volta's invention in 1800 of the voltaic pile, or battery. This device was a source of constant electric current and led to the subsequent age of electric power. In 1801 Volta gave a demonstration in Paris of his battery's operation before Napoleon, who made Volta a count and senator of the kingdom of Lombardy.

SIR WILLIAM HERSCHEL (1738–1822) AND CAROLINE HERSCHEL (1750–1848)

German-born British astronomer, the founder of sidereal astronomy for the systematic observation of the heavens; and his sister, who aided him.

William Herschel was born in Hannover, Hanover, where his father was an army musician. Following the same profession, the boy played in the band of the Hanoverian Guards. After the French occupation of Hanover in 1757 he escaped to England, where at first he earned a living by copying music. But he steadily improved his position by becoming a music teacher, performer, and composer, until in 1766 he was appointed organist of a fashionable chapel in Bath, the well-known spa. Caroline assisted her mother in the management of the household in Hannover until 1772, when William took her to Bath. There she trained and performed successfully as a singer. (Both she and William gave their last public musical performance in 1782.)

While working in Bath, William was introduced to the techniques of telescope construction. Not content to observe the nearby sun, moon, and planets, as did nearly all astronomers of his day, he was determined to study the distant celestial bodies as well. He realized that he would need telescopes with large mirrors to collect enough light – larger, in fact, than opticians could supply at reasonable cost – and was soon forced to grind his own mirrors. Later, his telescopes proved far superior even to those used at the Greenwich Observatory. He also made his own eyepieces, the strongest with a magnifying power of 6,450 times.

At Bath he was helped in his research by Caroline, who was his faithful assistant throughout much of his career. News of this extraordinary household began to spread in scientific circles. William made two preliminary telescopic surveys of the heavens. Then, in 1781, during his third and most complete survey of the night sky, he came upon an object that he realized was not an ordinary star. It proved to be the planet Uranus, the first planet to be discovered since prehistoric times. William became famous almost overnight. His friend Dr William Watson Jr introduced him to the Royal Society, which awarded him the Copley Medal for the discovery of Uranus, and elected him a fellow. At this time William was appointed as an astronomer to George III, and the Herschels moved to Datchet, near Windsor Castle.

Although he was 43 years old, William worked night after night to develop a "natural history" of the heavens. A fundamental problem for which Herschel's big telescopes were ideally suited concerned the nature of nebulae, which appear as luminous patches in the sky. Some astronomers thought they were nothing more than clusters of innumerable stars, the light of which blends to form a milky appearance. Others held that some nebulae were composed of a luminous fluid.

William's interest in nebulae developed in the winter of 1781–2, and he quickly found that his most powerful telescope could resolve into stars several nebulae that appeared "milky" to less well-equipped observers. He was convinced that, with more powerful instruments, other nebulae would eventually be resolved into individual stars. This encouraged him to argue in 1784 and 1785 that all nebulae were formed of stars and that there was no need to postulate the existence of a mysterious luminous fluid to explain the observed facts. By this reasoning,

William was led to hyphothesize the existence of what were later called "island universes" (now known as galaxies) of stars.

In 1785 he thus developed a cosmogony – a theory concerning the origin of the universe – and gave the first major example of the usefulness of stellar statistics, in that he could count the stars and interpret this data in terms of the extent in space of the galaxy's star system. Other astronomers, cut off from the evidence by the modest size of their telescopes and unwilling to follow William in his bold theorizing, could only look on with varying degrees of sympathy or scepticism.

In 1787 the Herschels moved to Old Windsor, and the following year to nearby Slough, where William spent the rest of his life. William's achievement, in a field in which he became a professional only in middle life, was made possible by his own total dedication and the selfless support of Caroline. Night after night, whenever the moon and weather permitted, he observed the sky with Caroline, who recorded his observations. On overcast nights, William would post a watchman to summon him if the clouds should break. In the daytime, Caroline often summarized the results of their work while he directed the construction of telescopes, many of which he sold to supplement their income. As her interest grew, she swept the heavens with a small Newtonian reflector and made her own astronomical discoveries.

William's grand concept of stellar organization received a jolt on 13 November 1790 when he observed a remarkable nebula, which he was forced to interpret as a central star surrounded by a cloud of "luminous fluid". This discovery contradicted his earlier views. Hitherto he had reasoned that many nebulae that he was unable to resolve (separate into distinct stars), even with his best telescopes, might be distant "island universes". He was able, however, to adapt his earlier theory to this new evidence by concluding that the central star he had observed was condensing out of the surrounding cloud under the forces of gravity. In 1811 he extended his cosmogony backwards in time to the stage when stars had not yet begun to form out of the fluid.

Herschel's labours through 20 years of systematic sweeps for nebulae (1783–1802) resulted in three catalogues listing 2,500 nebulae and star clusters that he substituted for the 100 or so milky patches previously known. He also catalogued 848 double stars – pairs of stars that appear close together in space – and measurements of the comparative brightness of stars. He observed that double stars did not occur by chance, as a result of random scattering of stars in space, but that they actually revolved about each other. His 70 published papers include not only studies of the motion of the solar system through space and the announcement in 1800 of the discovery of infrared rays but also a succession of detailed investigations of the planets and other members of the solar system.

In 1798 Caroline presented to the Royal Society an *Index to Flamsteed's Observations*, together with a catalogue of 560 stars omitted from the *British Catalogue* and a list of the errata in that publication. She returned to Hannover after William's death in 1822 and soon completed the cataloguing of 2,500 nebulae and many star clusters. In 1828 the Astronomical Society awarded her its gold medal for an unpublished revision and reorganization of their work.

ANTOINE-LAURENT LAVOISIER
(1743–1794)

Prominent French chemist and leading figure
in the eighteenth-century chemical revolution.

Lavoisier was the first child and only son of a wealthy bourgeois family living in Paris. As a youth he studied law, but the Paris law faculty made few demands on its students and Lavoisier was able to spend much of his three years as a law student attending public and private lectures on chemistry and physics and working under the tutelage of leading naturalists. Upon completing his legal studies, Lavoisier, like his father and his maternal grandfather before him, was admitted to the elite Order of Barristers. But, rather than practise law, he began pursuing scientific research that in 1768 gained him admission into France's foremost natural philosophy society, the Academy of Sciences in Paris.

Shortly before entering the Academy of Sciences, Lavoisier had received a considerable inheritance from his mother's estate, which he used to purchase an interest in a financial enterprise known as the General Farm. This was a partnership that had a contract with the royal government to collect certain sales and excise taxes, such as those on salt and tobacco. Lavoisier spent a considerable time as a Tax Farmer, and he was richly rewarded for his efforts. Although chemistry was Lavoisier's passion, throughout his life he devoted the majority of his time to financial and administrative affairs.

After being elected a junior member of the Academy of Sciences, Lavoisier began searching for a field of research in which he could distinguish himself. Chemists had long recognized that burning, like breathing, required air, and they also knew that iron rusts only upon exposure to air. Noting that burning gives off light and heat, that warm-blooded animals breathe, and that ores are turned into metals in a furnace, they concluded that fire was the key causal element behind these chemical reactions. The Enlightenment German chemist Georg Ernst Stahl (1660–1734) provided a well-regarded explanation of these phenomena.

Stahl hypothesized that a common "fiery substance" he called phlogiston was released during combustion, respiration, and calcination, and that it was absorbed when these processes were reversed. Although plausible, this theory raised a number of problems for those who wished to explain chemical reactions in terms of substances that could be isolated and measured. In the early stages of his research Lavoisier regarded the phlogiston theory as a useful hypothesis, but he sought ways either to solidify its firm experimental foundation or to replace it with an experimentally sound theory of combustion. In the end, his theory of oxygenation replaced the phlogiston hypothesis, but it took Lavoisier many years and considerable help from others to reach this goal.

The oxygen theory of combustion resulted from a demanding and sustained campaign to construct an experimentally grounded chemical theory of combustion, respiration, and calcination. The theory that emerged was in many respects a mirror image of the phlogiston theory, but gaining evidence to support the new theory involved more than merely demonstrating the errors and inadequacies of the previous theory. From the early 1770s until 1785, when the last important pieces of the theory fell into place, Lavoisier and his collaborators performed a wide range of experiments designed to advance many points on their research frontier.

Lavoisier's research in the early 1770s focused upon weight

gains and losses in calcination. It was known that when metals slowly changed into powders (calxes), as was observed in the rusting of iron, the calx actually weighed more than the original metal, whereas when the calx was "reduced" to a metal, a loss of weight occurred. The phlogiston theory did not account for these weight changes, for fire itself could not be isolated and weighed. Lavoisier hypothesized that it was probably the fixation and release of air, rather than fire, that caused the observed gains and losses in weight. This idea set the course of his research for the next decade.

In the process, he encountered related phenomena that had to be explained. Mineral acids, for instance, were made by roasting a mineral such as sulphur in fire and then mixing the resultant calx with water. In addition, new kinds of airs within the atmosphere were being discovered. British chemists made most of these discoveries, with Joseph Priestley leading the effort. And it was Priestley, despite his unrelenting adherence to the phlogiston theory, who ultimately helped Lavoisier unravel the mystery of oxygen. Priestley isolated oxygen in August 1774 after recognizing several properties that distinguished it from atmospheric air. At the same time, in Paris, Lavoisier and his colleagues were experimenting with a set of reactions identical to those that Priestley was studying, but they failed to notice the novel properties of the air they collected. Priestley visited Paris later that year and, at a dinner held in his honour at the Academy of Sciences, informed his French colleagues about the properties of this new air. Lavoisier, who was familiar with Priestley's research and held him in high regard, hurried back to his laboratory, repeated the experiment, and found that it produced precisely the kind of air he needed to complete his theory. He called the gas that was produced oxygen ("the generator of acids"). Isolating oxygen allowed him to explain both the quantitative and qualitative changes that occurred in combustion, respiration, and calcination.

In the canonical history of chemistry Lavoisier is celebrated as the leader of the eighteenth-century chemical revolution and consequently one of the founders of modern chemistry. He was indeed an indefatigable and skilful investigator; however, his experiments emphasized quantification and demonstration rather than yielding critical discoveries. Such an emphasis suited his determination to elevate chemistry to the level of a rigorous science.

Lavoisier was fortunate in having made his contributions to the chemical revolution before the disruptions of political revolution. By 1785 his new theory of combustion was gaining support, and the campaign to reconstruct chemistry according to its precepts began. One tactic to enhance the wide acceptance of his new theory was to propose a related method of naming chemical substances. In 1787 Lavoisier and three prominent colleagues published a new nomenclature of chemistry, which was soon widely accepted, thanks largely to Lavoisier's eminence and the cultural authority of Paris and the Academy of Sciences. Its fundamentals remain the method of chemical nomenclature in use today. Two years later Lavoisier published a programmatic *Traité élémentaire de chimie* ("Elementary Treatise on Chemistry") that described the precise methods chemists should employ when investigating, organizing, and explaining their subjects. It was a worthy culmination of a determined and largely successful programme to reinvent chemistry as a modern science.

When the French Revolution began in 1789, Lavoisier, like many other philosophically minded administrators, saw it as an opportunity to rationalize and improve the nation's politics and economy. Such optimism was soon tempered, however, by upheavals that put the very existence of the state at risk. Lavoisier, perhaps overvaluing the authority of science and the power of reason, continued to advise revolutionary governments on finance and other matters, and neither he nor his wife fled abroad when popular anger turned against those who had exercised power and enjoyed social privileges in the old regime. As the revolution became increasingly radical and those in command were driven to ruling by terror, Lavoisier continued to argue that the Academy of Sciences should be saved because its members were loyal and indispensable servants of the state. This rearguard action was unsuccessful, and he soon found himself imprisoned along with other members of the General Farm. The Republic was being purged of its royalist past. In May 1794 Lavoisier, his father-in-law, and 26 other Tax Farmers were guillotined. Acknowledging Lavoisier's scientific stature, his contemporary, Joseph-Louis Lagrange, commented, "It took them only an instant to cut off that head, and a hundred years may not produce another like it."

PIERRE-SIMON, MARQUIS DE LAPLACE (1749–1827)

French mathematician, astronomer, and physicist, best known for his investigations into the stability of the solar system.

Laplace was the son of a peasant farmer in Normandy, France. Little is known of his early life except that he quickly showed his mathematical ability at the military academy at Beaumont. In 1766 he entered the University of Caen, but he left for Paris the next year, apparently without taking a degree. He arrived with a letter of recommendation to the mathematician Jean Le Rond d'Alembert, who helped him secure a professorship at the École Militaire, where he taught from 1769 to 1776.

In 1773 he began his major lifework – applying Newtonian gravitation to the entire solar system – by taking up a particularly troublesome problem: why Jupiter's orbit appeared to be continuously shrinking while Saturn's continually expanded. The mutual gravitational interactions within the solar system were so complex that mathematical solution seemed impossible; indeed, Newton (1642–1727) had concluded that divine intervention was periodically required to preserve the system in equilibrium. Laplace announced the invariability of planetary mean motions (average angular velocity). This discovery in 1773 – the first and most important step in establishing the stability of the solar system – was the most significant advance in physical astronomy since Newton. It won Laplace associate membership in the French Academy of Sciences the same year.

Applying quantitative methods to a comparison of living and non-living systems, in 1780 Laplace and the chemist Antoine-Laurent Lavoisier, with the aid of an ice calorimeter that they had invented, showed respiration to be a form of combustion. Returning to his astronomical investigations with an examination of the entire subject of planetary perturbations – mutual gravitational effects – Laplace in 1786 proved that the eccentricities and inclinations of planetary orbits to each other will always remain small, constant, and self-correcting. The effects of perturbations were therefore conservative and periodic, not cumulative and disruptive.

During 1784–5 Laplace worked on the subject of attraction between spheroids; in this work the potential function of later physics can be recognized for the first time. Laplace explored the problem of the attraction of any spheroid upon a particle situated outside or upon its surface. Through his discovery that the attractive force of a mass upon a particle, regardless of direction, can be obtained directly by differentiating a single function, Laplace laid the mathematical foundation for the scientific study of heat, magnetism, and electricity.

He removed the last apparent anomaly from the theoretical description of the solar system in 1787 with the announcement that lunar acceleration depends on the eccentricity of the Earth's orbit. Although the mean motion of the moon around the Earth depends mainly on the gravitational attraction between them, it is slightly diminished by the pull of the sun on the moon. This solar action depends, however, on changes in the eccentricity of the Earth's orbit resulting from perturbations by the other planets. As a result, the moon's mean motion is accelerated as long as the Earth's orbit tends to become more circular, but, when the reverse occurs, this motion is retarded. Laplace concluded that the inequality is therefore not truly cumulative but is of a period running into millions of years. The last threat of instability thus disappeared from the theoretical description of the solar system.

In 1796 Laplace published *Exposition du système du monde* ("The System of the World"), a semi-popular treatment of his work in celestial mechanics and a model of French prose. The book included his "nebular hypothesis" – attributing the origin of the solar system to the cooling and contracting of a gaseous nebula – which strongly influenced future thought on planetary origin. His *Traité de mécanique céleste* ("Celestial Mechanics"), appearing in five volumes between 1798 and 1827, summarized the results obtained by his mathematical development and application of the law of gravitation. He offered a complete mechanical interpretation of the solar system by devising methods for calculating the motions of the planets, their satellites and their perturbations, including the resolution of tidal problems. The book made him a celebrity.

Probably because he did not hold strong political views and was not a member of the aristocracy, Laplace escaped imprisonment and execution during the French Revolution. He was president of the Board of Longitude, aided in the organization of the metric system, helped found the scientific Society of Arcueil, and was created a marquis. He served for six weeks as minister of the interior under Napoleon, who famously reminisced that Laplace "carried the spirit of the infinitesimal into administration."

EDWARD JENNER (1749–1823)

English surgeon and discoverer
of the vaccination for smallpox.

Jenner was born in Berkeley, Gloucestershire. He was a country youth; the son of a clergyman. He attended grammar school and at the age of 13 was apprenticed to a nearby surgeon. In the following eight years Jenner acquired a sound knowledge of medical and surgical practice. On completing his apprenticeship at the age of 21, he went to London and became the house pupil of John Hunter, one of the most prominent surgeons in London. Even more importantly, however, he was an anatomist, biologist, and experimentalist of the first rank. In addition to his training and experience in biology, Jenner made progress in clinical surgery. After studying in London from 1770 to 1773, he returned to country practice in Berkeley and enjoyed substantial success. He was capable, skilful, and popular.

Smallpox was widespread in the eighteenth century, and occasional outbreaks of particular intensity resulted in a very high death rate. The disease respected no social class, and disfigurement was not uncommon in patients who recovered. The only means of combating smallpox was a primitive form of vaccination called variolation – intentionally infecting a healthy person with the "matter" taken from a patient sick with a mild attack of the disease. The practice, which originated in China and India, was based on two distinct concepts: firstly, that one attack of smallpox effectively protected against any subsequent attack, and secondly, that a person deliberately infected with a mild case of the disease would safely acquire such protection. Unfortunately, the transmitted disease did not always remain mild, and mortality sometimes occurred. Furthermore, the inoculated person could disseminate the disease to others and thus act as a focus of infection.

Jenner had been impressed by the fact that a person who had suffered an attack of cowpox – a relatively harmless disease that could be contracted from cattle – could not become infected, whether by accidental or intentional exposure, to smallpox. Pondering this phenomenon, he concluded that cowpox not only protected against smallpox but could be transmitted from one person to another as a deliberate mechanism of protection.

In May 1796 Jenner found a young dairymaid, Sarah Nelmes, who had fresh cowpox lesions on her hand. On 14 May, using matter from Sarah's lesions, he inoculated an eight-year-old boy, James Phipps, who had never had smallpox. Phipps became slightly ill over the course of the next nine days but was well on the tenth. On 1 July Jenner inoculated the boy again, this time with smallpox matter. No disease developed: the protection was complete. In 1798 Jenner, having added further cases, published privately a slender book entitled *An Inquiry into the Causes and Effects of the Variolae Vaccinae.*

The reaction to the publication was not immediately favourable. Jenner went to London seeking volunteers for vaccination but, in a stay of three months, was not successful. Vaccination rapidly proved its value, however, and Jenner became intensely active promoting it. The procedure spread rapidly to America and the rest of Europe and was soon carried around the world.

Despite errors and occasional chicanery, the death rate from smallpox plunged. Jenner received worldwide recognition and many honours, but he made no attempt to enrich himself through his discovery, and actually devoted so much time to the cause of vaccination that his private practice and personal affairs suffered severely. Parliament voted him a sum of £10,000 in 1802 and a further sum of £20,000 in 1806.

GEORGES, BARON CUVIER (1769–1832)

French zoologist and statesman, who established the sciences of comparative anatomy and palcontology.

Cuvier was born in Montbéliard, France. In 1784–8 he attended the Académie Caroline (Karlsschule) in Stuttgart, Germany, where he studied comparative anatomy and learned to dissect. After graduation Cuvier served from 1788 to 1795 as a tutor, during which time he wrote original studies of marine invertebrates, particularly the molluscs. Soon he joined the staff of the Museum of Natural History in Paris.

Cuvier refused an invitation to become a naturalist on Napoleon's expedition to Egypt in 1798–1801, preferring to remain at the museum to continue his research in comparative anatomy.

His first result, in 1797, was *Tableau élémentaire de l'histoire naturelle des animaux* ("Elementary Survey of the Natural History of Animals"), a popular work based on his lectures. In 1800–05 he published his *Leçons d'anatomie comparée* ("Lessons on Comparative Anatomy"). In this work, based also on his lectures at the museum, he put forward his principle of the "correlation of parts", according to which the anatomical structure of every organ is functionally related to all other organs in the body of an animal, and the functional and structural characteristics of organs result from their interaction with their environment. Moreover, according to Cuvier, the functions and habits of an animal determine its anatomical form.

Cuvier also argued that the anatomical characteristics distinguishing groups of animals are evidence that species had not changed since the Creation. Each species is so well coordinated, functionally and structurally, that it could not survive significant change. He further maintained that each species was created for its own special purpose and each organ for its special function. In denying evolution, Cuvier disagreed with the views of his colleague Jean-Baptiste Lamarck, who published his theory of evolution in 1809, and eventually also with Étienne Geoffroy Saint-Hilaire, who in 1825 published evidence concerning the evolution of crocodiles.

Cuvier advanced rapidly. While continuing his zoological work at the museum he brought about major reforms in education. He served as imperial inspector of public instruction and assisted in the establishment of French provincial universities. For these services he was granted the title "chevalier" in 1811. He also wrote the *Rapport historique sur les progrès des sciences naturelles depuis 1789* ("Historical Report on the Progress of the Natural Sciences Since 1789"), published in 1810. These publications are lucid expositions of the European science of his time.

Meanwhile, Cuvier also applied his views on the correlation of parts to a systematic study of fossils that he had excavated. He reconstructed complete skeletons of unknown fossil quadrupeds, which constituted astonishing new evidence that whole species of animals had become extinct. Furthermore, he discerned a remarkable sequence in the creatures he exhumed. The deeper, more remote strata contained animal remains – giant salamanders, flying reptiles, and extinct elephants – that were far less similar to animals now living than those found in the more recent strata. He summarized his conclusions, first in 1812 in his *Recherches sur les ossements fossiles de quadrupèdes* ("Researches on the Bones of Fossil Vertebrates"), as well as in the expansion of this essay in book form in 1825, *Discours sur les révolutions de la surface du globe* ("Discourse on the Revolutions of the Globe").

Cuvier assumed a relatively short time span for the Earth but was impressed by the vast changes that undoubtedly had occurred in its geologic past. His work gave new prestige to the old concept of catastrophism, according to which a series of "revolutions", or catastrophes – sudden land upheavals and floods – had destroyed entire species of organisms and carved out the present features of the Earth. He believed that the area laid waste by these spectacular paroxysms, of which Noah's flood was the most recent and dramatic, was sometimes repopulated by migration of animals from an area that had been spared. Catastrophism remained a major geologic doctrine until it was shown that slow changes over long periods of time could explain the Earth's features.

In 1817 Cuvier also published *Le Règne animal distribué d'après son organisation* ("The Animal Kingdom, Distributed According to its Organization"), which, with its many subsequent editions, was a significant advance over the systems of classification established by Linnaeus. Cuvier showed that animals possessed so many diverse anatomical traits that they could not be arranged in a single linear system. Instead, he arranged animals into four large groups (vertebrates, molluscs, articulates, and radiates), each of which had a special type of anatomical organization. All animals within the same group were classified together, as he believed they were all modifications of one particular anatomical type.

Cuvier's lifework may be considered as marking a transition between the eighteenth-century view of nature and the view that emerged in the last half of the nineteenth century as a result of the doctrine of evolution. By rejecting the eighteenth-century method of arranging animals in a continuous series, in favour of classifying them in four separate groups, he raised the key question of why animals were anatomically different. Although Cuvier's doctrine of catastrophism did not last, he did set the science of palaeontology on a firm, empirical foundation. He did this by introducing fossils into zoological classification, showing the progressive relation between rock strata and their fossil remains, and by demonstrating, in his comparative anatomy and his reconstructions of fossil skeletons, the importance of functional and anatomical relationships.

ALEXANDER VON HUMBOLDT
(1769–1859)

German naturalist and explorer who was
a major figure in the classical period of
physical geography and biogeography.

Humboldt was born in Berlin, the son of an officer in the army of Frederick the Great. He was privately educated; instruction in political history and economics was added to the usual courses in classics, languages, and mathematics, as his mother intended him to be qualified for a high public position. After a year spent at the University of Göttingen, from 1789 to 1790, Humboldt became particularly interested in mineralogy and geology and decided to obtain a thorough training in these subjects by joining the School of Mines in Freiberg, Saxony, the first such establishment. There, buttressed by a prodigious memory and driven by an unending thirst for knowledge, he began to develop his enormous capacity for work. After a morning spent underground in the mines, he attended classes for five or six hours in the afternoon, and in the evening – pursuing his passionate interest in botany – he scoured the country for plants.

Humboldt left Freiberg in 1792 after two years of intensive study but without taking a degree. A month later he obtained an appointment in the Mining Department of the Prussian government and departed for the remote Fichtel Mountains in the Margraviate of Ansbach-Bayreuth, which had only recently come into the possession of the Prussian kings. Here Humboldt came into his own: he travelled untiringly from one mine to the next, reorganizing the partly deserted and totally neglected pits, which produced mainly gold and copper. He supervised all mining activities, invented a safety lamp, and established, with his own funds, a technical school for young miners. Yet he did not intend to make mining his career.

The conviction had grown in Humboldt that his real aim in life was scientific exploration, and in 1797 he resigned from his post to acquire with great single-mindedness a thorough knowledge of the systems of geodetic, meteorological, and geomagnetic measurements. He obtained permission from the Spanish government to visit the Spanish colonies in Central and South America. These colonies were then accessible only to Spanish officials and the Roman Catholic mission. Completely shut off from the outside world, they offered enormous possibilities to a scientific explorer. Humboldt's social standing assured him of access to official circles, and in the Spanish prime minister Mariano de Urquijo he found an enlightened man who supported his application to the king for a royal permit.

In the summer of 1799 he set sail from Marseille accompanied by the French botanist Aimé Bonpland, whom he had met in Paris, then the liveliest scientific centre in Europe. Humboldt and Bonpland spent five years, from 1799 to 1804, in Central and South America, covering more than 9,650 km (6,000 miles) on foot, on horseback, and in canoes. It was a life of great physical exertion and serious deprivation. Starting from Caracas, they travelled south through grasslands and scrublands until they reached the banks of the Apure, a tributary of the Orinoco River. They continued their journey on the river by canoe as far as the Orinoco. Following its course and that of the Casiquiare, they proved that the Casiquiare River formed a connection between the vast river systems of the Amazon and the Orinoco. For three months Humboldt and Bonpland moved through dense tropical forests, tormented by clouds of mosquitoes and stifled by the humid heat. Their provisions were soon destroyed by insects and rain, and the lack of food finally drove them to subsist on ground-up wild cacao beans and river water. Yet both travellers, buoyed up by the excitement provided by the new and overwhelming impressions, remained healthy and in the best of spirits until their return to civilization, when they succumbed to a severe bout of fever.

After a short stay in Cuba, Humboldt and Bonpland returned to South America for an extensive exploration of the Andes. From Bogotá to Trujillo, Peru, they wandered over the Andean highlands, following a route now traversed by the Pan-American Highway – in their time a series of steep, rocky, and often very narrow paths. They climbed a number of peaks, including all the volcanoes in the surroundings of Quito, Ecuador. Humboldt's ascent of Chimborazo, a peak of 6,310 metres (20,702 feet) to a height of 5,878 metres (19,286 feet), but short of the summit, remained a world mountain-climbing record for nearly 30 years.

All these achievements were carried out without the help of modern mountaineering equipment – ropes, crampons, or oxygen supplies – and Humboldt and Bonpland suffered badly from mountain sickness. But Humboldt turned his discomfort to advantage: he became the first person to ascribe mountain sickness to lack of oxygen in the rarefied air of great heights. He also studied the oceanic current off the west coast of South America – which was originally named after him but is now known as the Peru Current. When the pair arrived, worn and footsore, in Quito, Humboldt had no difficulty in assuming the role of courtier and man of the world when he was received by the Viceroy and the leaders of Spanish society. In the spring of 1803 the two travellers sailed from Guayaquil to Acapulco, Mexico, where they spent the last year of their expedition in a close study of this most developed and highly civilized part of the Spanish colonies. After a short stay in the United States, where Humboldt was received by President Jefferson, they sailed for France.

Humboldt and Bonpland returned to Europe with an immense amount of information. In addition to a vast collection of new plants, there were determinations of longitudes and latitudes, measurements of the components of the Earth's geomagnetic field, and daily observations of temperatures and barometric pressure, as well as statistical data on the social and economic conditions of Mexico.

The years from 1804 to 1827 Humboldt devoted to publication of the data accumulated on the South American expedition. With the exception of brief visits to Berlin, he lived in Paris during this period. There he found not only collaborators among the French scientists – the greatest of his time – but engravers for his maps and illustrations, and publishers for printing the 30 volumes into which the scientific results of the expedition were distilled. Of great importance were the meteorological data, with an emphasis on mean daily and nightly temperatures, and Humboldt's representation on weather maps of isotherms (lines connecting points with the same mean temperature) and isobars (lines connecting points with the same barometric pressure for a given time or period) – all of which helped lay the foundation for the science of comparative climatology.

Even more important were his pioneering studies on the relationship between a region's geography and its flora and fauna, and, above all, the conclusions he drew from his study of the Andean volcanoes concerning the role played by eruptive forces and the metamorphosis of rock in the history and ongoing development of the Earth's crust. These conclusions disproved once and for all the hypothesis of the so-called Neptunists, who held that the surface of the Earth had been totally formed by sedimentation from a liquid state.

Lastly, his *Political Essay on the Kingdom of New Spain* contained a wealth of material on the geography and geology of Mexico, including descriptions of its political, social, and economic conditions, and also extensive population statistics. Humboldt's impassioned outcry in this work against the inhumanities of slavery remained unheard, but his descriptions of the Mexican silver mines led to widespread investment of English capital and mining expertise there.

The happy years in Paris came to an end in 1827. Humboldt's means were by then almost completely exhausted; unable to maintain his financial independence, he had to return to Berlin, where the King impatiently demanded his presence at court. Until a few years before his death,

Humboldt served as a tutor to the Crown Prince, as a member of the privy council, and as a court chamberlain.

His enthusiasm for the popularization of science prompted him to give a course on physical geography to the professors and students of all faculties of the University of Berlin, part of which he repeated in a public lecture to an audience of more than 1,000. In the autumn of the same year, 1828, he also organized in Berlin one of the first international scientific conferences. Such large gatherings of possibly liberal-minded people were frowned upon by governments in the wake of the Napoleonic Wars and the attendant rise of democratic expectations, and it is a tribute to Humboldt's adroitness that he was able to overcome the misgivings of official Prussian circles.

During the last 25 years of his life, Humboldt was chiefly occupied with writing *Kosmos*, one of the most ambitious scientific works ever published. Four volumes appeared during his lifetime. *Kosmos* gives a generally comprehensible account of the structure of the universe as then known, at the same time communicating the scientist's excitement and aesthetic enjoyment at his discoveries. While still working on the fifth volume of *Kosmos*, Humboldt died in his 90th year.

SOPHIE GERMAIN (1776–1831)

French mathematician who contributed
notably to the study of acoustics, elasticity,
and the theory of numbers.

Germain was born in Paris. As a girl she read widely in her father's library; later, using the pseudonym of M. Le Blanc, she managed to obtain lecture notes for courses from the newly

organized École Polytechnique in Paris. It was through the
École Polytechnique that she met the mathematician Joseph-
Louis Lagrange, who remained a strong source of support and
encouragement to her for several years. Germain's early work
was in number theory, her interest having been stimulated by
Adrien-Marie Legendre's *Théorie des nombres* (1789) and by
Carl Friedrich Gauss's *Disquisitiones Arithmeticae* (1801).
This subject occupied her throughout her life and eventually
provided her most significant result. In 1804 she initiated a
correspondence with Gauss under her male pseudonym. Gauss
only learned of her true identity when Germain, fearing for
Gauss's safety as a result of the French occupation of Hannover
in 1807, asked a family friend in the French army to ascertain
his whereabouts and ensure that he would not be ill-treated.

In 1809 the French Academy of Sciences offered a prize for a
mathematical account of the phenomena exhibited in experiments on vibrating plates conducted by the German physicist
Ernst F.F. Chladni. In 1811 Germain submitted an anonymous
memoir, but the prize was not awarded. The competition was
reopened twice more, once in 1813 and again in 1816, and
Germain submitted a memoir on each occasion. Her third
memoir, with which she finally won the prize, treated vibra-
tions of general curved as well as plane surfaces and was
published privately in 1821. During the 1820s she worked on
generalizations of her research but, isolated from the academic
community on account of her gender and thus largely unaware
of new developments taking place in the theory of elasticity,
she made little real progress. In 1816 Germain met Joseph
Fourier, whose friendship and position in the Academy helped
her to participate more fully in Parisian scientific life, but his
reservations about her work on elasticity eventually led him to
distance himself from her professionally, although they remained close friends.

Meanwhile Germain had actively revived her interest in number theory, and in 1819 she wrote to Gauss outlining her strategy for a general solution to Fermat's last theorem, which states that there is no solution for the equation $x^n + y^n = z^n$ if n is an integer greater than 2 and x, y, and z are non-zero integers. She proved the special case in which x, y, z, and n are all relatively prime (have no common divisor except for 1) and n is a prime smaller than 100, although she did not publish her work. Her result first appeared in 1825 in a supplement to the second edition of Legendre's *Théorie des nombres*. She corresponded extensively with Legendre, and her method formed the basis for his proof of the theorem for the case $n = 5$. The theorem was proved for all cases by the English mathematician Andrew Wiles in 1995.

CARL FRIEDRICH GAUSS (1777–1855)

German mathematician and astronomer,
generally regarded as one of the
greatest mathematicians of all time.

Born in Brunswick, Germany, Gauss was the only child of poor parents. He was rare among mathematicians in that he was a calculating prodigy, and he retained the ability to do elaborate calculations in his head for most of his life. Impressed by this ability and by his gift for languages, his teachers and his devoted mother recommended him to the Duke of Brunswick in 1791, who granted him financial assistance to continue his education locally and then to study mathematics at the University of Göttingen from 1795 to 1798. Gauss's pioneering work gradually established him as the era's pre-eminent mathematician, first in the German-speaking world and then farther afield.

Gauss's first significant discovery, in 1792, was that a regular polygon of 17 sides can be constructed by ruler and compass alone. Its significance lies not in the result but in the proof, which rested on a profound analysis of the factorization of polynomial equations and opened the door to later ideas of Galois theory. Gauss's doctoral thesis of 1797 gave a proof of the fundamental theorem of algebra: that every polynomial equation with real or complex coefficients has as many roots (solutions) as its degree (the highest power of the variable). Gauss's proof, though not wholly convincing, was remarkable for its critique of earlier attempts. He later gave three more proofs of this major result, the last on the 50th anniversary of the first, which shows the importance he attached to the topic.

Gauss's recognition as a truly remarkable talent, however, resulted from two major publications in 1801. Foremost was his publication of the first systematic textbook on algebraic number theory, *Disquisitiones Arithmeticae*. The book begins with the first account of modular arithmetic, gives a thorough account of the solutions of quadratic polynomials in two variables in integers, and ends with the theory of factorization mentioned above. This choice of topics and its natural generalizations set the agenda in number theory for much of the nineteenth century.

The second publication, *Theoria motus corporum coelestium* ("Theory of the Movement of Heavenly Bodies"), referred to his rediscovery of the asteroid Ceres. Its original discovery, by the Italian astronomer Giuseppe Piazzi in 1800, had caused a sensation, but the asteroid vanished behind the sun before enough observations could be taken to calculate its orbit with sufficient accuracy to know where it would reappear. Many astronomers competed for the honour of finding it again, but Gauss won. His success rested on a novel method for dealing with errors in observations, today called the method of least squares. Thereafter Gauss worked for many years as an astronomer and published a major work on the computation of orbits.

Gauss then accepted the challenge of surveying the territory of Hanover, and he was often out in the field in charge of the observations. The project, which lasted from 1818 to 1832, encountered numerous difficulties, but it led to a number of advancements. One was Gauss's invention of the heliotrope (an instrument that reflects the sun's rays in a focused beam that can be observed from several miles away), which improved the accuracy of the observations. Another was his discovery of a way of formulating the concept of the curvature of a surface. Gauss showed that there is an intrinsic measure of curvature that is not altered if the surface is bent without being stretched. He then published works on number theory, the mathematical theory of map construction, and many other subjects.

In the 1830s Gauss became interested in terrestrial magnetism and participated in the first worldwide survey of the Earth's magnetic field (to measure it, he invented the magnetometer). With his Göttingen colleague, the physicist Wilhelm Weber, he made the first electric telegraph, but a certain parochialism prevented him from pursuing the invention energetically. Instead, he drew important mathematical consequences from this work for what is today called potential theory, an important branch of mathematical physics arising in the study of electromagnetism and gravitation.

Gauss also wrote on cartography, the theory of map projections. For his study of angle-preserving maps he was awarded the prize of the Danish Academy of Sciences in 1823. This work came close to suggesting that complex functions of a complex variable are generally angle preserving, but Gauss stopped short of making that fundamental insight explicit, leaving it for Bernhard Riemann (1826–66), who had a deep appreciation of Gauss's work. Gauss also had other unpublished insights into the nature of complex functions and their integrals, some of which he divulged to friends. In fact, Gauss often withheld publication of his discoveries.

In 1812 he published an account of an interesting infinite series, called the hypergeometric series, and he wrote but did not publish an account of the differential equation that it satisfies. He showed that the series can be used to define many familiar and many new functions. This was a major breakthrough, because, as Gauss had discovered in the 1790s, the contemporary theory of complex integrals was utterly inadequate for the task. When some of this theory was published by the Norwegian Niels Abel and the German Carl Jacobi in about 1830, Gauss commented to a friend that Abel had come one third of the way. This was accurate, but it was telling of Gauss's personality that he still withheld publication.

After Gauss's death the discovery of so many novel ideas among his unpublished papers extended his influence well into the remainder of the century. Acceptance of non-Euclidean geometry had not come with the original publications of the Russian Nicolay Lobachevsky and the Hungarian János Bolyai in 1829 and 1831 respectively, but came instead with the almost simultaneous publication of Riemann's general ideas about geometry, the explicit and rigorous account of it by the Italian Eugenio Beltrami, and Gauss's private notes and correspondence.

SIR HUMPHRY DAVY, BARONET (1778–1829)

English chemist who discovered several chemical elements and compounds and invented the miner's safety lamp.

Davy was educated at the grammar school in Penzance, Cornwall, and, in 1793, at Truro. In 1795, a year after the death of his father Robert, he was apprenticed to a surgeon and apothecary, and he hoped eventually to qualify in medicine. In 1797 he was befriended by Davies Giddy (later Gilbert; president of the Royal Society 1827–30), who offered him the use of his library at Tredrea in Cornwall and took him to a well-equipped chemistry laboratory. There Davy formed strongly independent views on topics of the moment, such as the nature of heat, light, and electricity, and the chemical and physical doctrines of the French chemist Antoine-Laurent Lavoisier.

On Gilbert's recommendation, Davy was appointed in 1798 to chemical superintendent of the Pneumatic Institution, founded at Clifton to inquire into the possible therapeutic uses of various gases. Davy attacked the problem with characteristic enthusiasm, evincing an outstanding talent for experimental inquiry. He investigated the composition of the oxides and acids of nitrogen, as well as ammonia, and persuaded his scientific and literary friends, including Samuel Taylor Coleridge, Robert Southey, and P.M. Roget, to report the effects of inhaling nitrous oxide. He nearly lost his own life inhaling water gas, a mixture of hydrogen and carbon monoxide sometimes used as fuel. The account of his work, published as *Researches, Chemical and Philosophical* (1800), immediately established his reputation, and he was invited to lecture at the newly founded Royal Institution of Great Britain in London, where he moved in 1801.

His carefully prepared and rehearsed lectures rapidly became important social functions and added greatly to the prestige of science and the institution. In 1802 he became professor of chemistry. His duties included a special study of leather tanning: he found catechu, the extract of a tropical plant, to be as effective as and cheaper than the usual oak extracts, and his published account was long used as a tanner's guide. In 1803 he was admitted a fellow of the Royal Society and an honorary member of the Dublin Society and delivered the first of an annual series of lectures before the Board of Agriculture. This led to his *Elements of Agricultural Chemistry* (1813), the only systematic work on the subject available for many years. For his researches on voltaic cells, tanning, and mineral analysis, he received the Copley Medal in 1805. He was elected secretary of the Royal Society in 1807.

Davy early concluded that the production of electricity in simple electrolytic cells resulted from chemical action. He therefore reasoned that electrolysis, the interactions of electric currents with chemical compounds, offered the most likely means of decomposing all substances to their elements. These views were explained in 1806 in his lecture "On Some Chemical Agencies of Electricity", for which, despite the fact that England and France were at war, he received the Napoleon Prize from the Institut de France (1807). This work led directly to the isolation of sodium and potassium from their compounds (1807) and of the alkaline-earth metals from theirs (1808). Davy also discovered boron (by heating borax with potassium), hydrogen telluride, and hydrogen phosphide (phosphine). He showed the correct relation of chlorine to hydrochloric acid and the untenability of the earlier name (oxymuriatic acid) for chlorine.

In 1812 Davy was knighted by the Prince Regent. That year he also published the first part of the *Elements of Chemical Philosophy*, which contained much of his own work, and delivered a farewell lecture to members of the Royal Institution. His last important act at the Royal Institution, of which he remained honorary professor, was to interview the young Michael Faraday, later to become one of England's great scientists, who became laboratory assistant there in 1813. Davy was created a baronet in 1818.

During his later years Davy studied the conditions under which mixtures of firedamp (a flammable gas found in coal mines) and air explode. This work, for the Society for Preventing Accidents in Coal Mines, led to his invention of the miner's safety lamp. In 1820 Davy became president of the Royal Society, a position he held until 1827. Davy's health was by then failing rapidly; in 1827 he departed for Europe and, in the summer of that year, was forced to resign the presidency of the Royal Society. After a last, short visit to England he retired to Italy, settling at Rome in February 1829 – "a ruin amongst ruins". Though partly paralyzed through stroke, he spent his last months writing a series of dialogues, published posthumously as *Consolations in Travel, or the Last Days of a Philosopher* (1830).

JÖNS JACOB BERZELIUS (1779–1848)

One of the founders of modern chemistry, who is especially noted for his determination of atomic weights and the development of modern chemical symbols.

Berzelius was born near Linköping, Sweden. He studied medicine at Uppsala University from 1796 to 1802, and from 1807 to 1832 he served as a professor of medicine and pharmacy at the Karolinska Institute. He became a member of the Royal Swedish Academy of Sciences in 1808, serving from 1818 as its principal functionary, the perpetual secretary. He was an early Swedish supporter of the new chemistry proposed a generation earlier by the renowned French chemist Antoine-Laurent Lavoisier, and he remained a forceful exponent of Enlightenment science and progressive politics.

Berzelius is best known for his system of electrochemical dualism. The electrical battery, invented in 1800 by Alessandro Volta (1745–1827) and known as the voltaic pile, provided the first experimental source of current electricity. In 1803 Berzelius demonstrated, as did the English chemist Humphry Davy at a slightly later date, the power of the voltaic pile to decompose chemicals into pairs of electrically opposite constituents. For Berzelius, all chemical compounds contained two electrically opposing constituents: the acidic, or electronegative, and the basic, or electropositive. His generalization elevated bases from their formerly passive role as mere substrates upon which acids reacted to form salts, to substances having characteristic properties opposite to those of acids. He also generalized about the electrochemical dualism of other substances. According to Berzelius, all chemicals – whether natural or artificial; mineral or organic – could be distinguished and specified qualitatively by identifying their electrically opposing constituents.

In addition to his qualitative specification of chemicals, Berzelius investigated their quantitative relationships. As early as 1806 he began to prepare an up-to-date Swedish chemistry textbook and read widely on the subject of chemical combination. Finding little information on the subject, he decided to undertake further investigations, which in turn focused his attention upon inorganic chemistry, and in around 1808 he launched what became a vast and enduring programme in the laboratory analysis of inorganic matter. To this end, he created most of his apparatuses and prepared his own reagents. Through precise experimental trials, supported by extraordinary interpretative acumen, he established the atomic weights of the elements, the formulae of their oxides, sulphides, and salts, and the formulae of virtually all known inorganic compounds, many of which he was the first to prepare or characterize.

Berzelius's experiments led to a more complete depiction of the principles of chemical combining proportions, an area of investigation that the German chemist Jeremias Benjamin Richter named "stoichiometry" in 1792. By showing how compounds conformed to the laws of constant, multiple, and equivalent proportions as well as to a series of semiempirical rules devised to cover specific classes of compounds, Berzelius established the quantitative specificity by which substances combined. He reported his analytical results in a series of famous publications, most prominently his "Essai sur la théorie des proportions chimiques et sur l'influence chimique de l'électricité" ("Essay on the Theory of Chemical Proportions and on the Chemical Influence of Electricity"; 1819), and the atomic weight tables that appeared in the 1826 German translation of his *Lärbok i kemien* ("Textbook of Chemistry"). He continued his analytical work until 1844, reporting in specialized articles and new editions of his textbook both new results, such as his extensive analysis of the compounds of the platinum metals in 1827–8, and refinements of his earlier experimental findings.

The project of specifying substances had several important consequences. In order to establish and display the laws of stoichiometry, Berzelius invented and perfected more exacting standards and techniques of analysis. His generalization of the older acid/base chemistry led him to extend the chemical nomenclature that Lavoisier had introduced to cover the bases (mostly metallic oxides): a change that allowed Berzelius to name any compound consistently with Lavoisier's chemistry.

The project of specifying substances also led Berzelius to develop a new system of notation that could portray the composition of any compound both qualitatively (by showing its electrochemically opposing ingredients) and quantitatively (by showing the proportions in which the ingredients were united). His system abbreviated the Latin names of the elements to one or two letters and applied superscripts (subscripts in modern notation) to designate the number of atoms of each element present in both the acidic and basic ingredient.

Berzelius applied his analytical method to two primary areas, mineralogy and organic chemistry. Cultivated in Sweden for its industrial utility, mineralogy had long stimulated Berzelius's analytical interest. Berzelius himself discovered several new elements, including cerium (1803) and thorium (1828), in samples of naturally occurring minerals, and his students discovered lithium, vanadium, lanthanum, didymium (later resolved into praseodymium and neodymium), erbium (later resolved into erbium, ytterbium, scandium, holmium, and thulium), and terbium. Berzelius also discovered selenium

(1818), though this element was isolated in the mud resulting from the manufacture of sulphuric acid rather than from a mineral sample. Berzelius's interest in mineralogy also fostered his analysis and preparation of new compounds of these and other elements.

Berzelius had a profound influence on chemistry, stemming in part from his substantial achievements and in part from his ability to enhance and project his authority. Throughout his life he cultivated professional relationships in diverse ways. He trained a number of leading students at home and abroad and also maintained a vast correspondence with professional colleagues. He was equally industrious in disseminating information about his ideas, methods, and results. To this end he published his scientific articles in French, German, and English, and frequently revised his *Textbook of Chemistry* in French and German editions that were often prepared with the help of current or former students. Finally, as perpetual secretary of the Royal Swedish Academy of Sciences, he issued annual reports from 1821 to 1848 (in Swedish, German, and French) on the progress of science. These reports not only announced his major findings but also offered Olympian pronouncements that were eagerly awaited, sometimes feared, but long highly respected.

JOHN JAMES AUDUBON (1785–1851)

Ornithologist, artist, and naturalist who
became particularly well known for his drawings
and paintings of North American birds.

Audubon was the illegitimate son of a French merchant, planter, and slave trader and a Creole woman of Saint-Domingue.

He and his illegitimate half-sister (who was also born in the West Indies) were legalized by adoption in 1794, five years after their father returned to France. The young Audubon developed an interest in drawing birds during his boyhood in France. At the age of 18 he was sent to the United States in order to avoid conscription and to enter business, and began his study of North American birds at that time. By 1820, after several failed business ventures, Audubon had begun to take what jobs he could to provide a living and to concentrate on his steadily growing interest in drawing birds; he worked for a time as a taxidermist, later making portraits and teaching drawing.

By 1824 he began to consider publication of his bird drawings, but he was advised to seek a publisher in Europe, where he would find better engravers and greater interest in his subject. In 1826 he went to Europe in search of patrons and a publisher. He was well received in Edinburgh and, after the king subscribed for his books, in London too. The engraver Robert Havell of London undertook publication of his illustrations as *The Birds of America* (1827–38), in four volumes with 435 hand-coloured plates. William MacGillivray helped write the accompanying text, *Ornithological Biography* (1831–9), in five volumes, and *A Synopsis of the Birds of North America*, (1839), which serves as an index.

Until 1839 Audubon divided his time between Europe and the United States, gathering material, completing illustrations, and financing publication through subscription. His reputation established, he then settled in New York City and prepared a smaller edition of his *Birds of America* (1840–4), in seven volumes; and a new work, *Viviparous Quadrupeds of North America* (1845–8), in three volumes with 150 plates, along with the accompanying text (1846–53) in three volumes, completed with the aid of his sons and the naturalist John Bachman.

Critics of Audubon's work have pointed to certain fanciful (or even impossible) poses and inaccurate details, but few argue with their excellence as art. To many, Audubon's work far surpasses that of his contemporary (and more scientific) fellow ornithologist, Alexander Wilson.

MICHAEL FARADAY (1791–1867)

English physicist and chemist whose many experiments contributed greatly to the understanding of electromagnetism.

Faraday was born in the country village of Newington, Surrey, now a part of south London. His father was a blacksmith who had migrated from the north of England earlier in 1791 to look for work. Faraday received only the rudiments of an education, learning to read, write, and cipher in a church Sunday school. At an early age he began to earn money by delivering newspapers for a book dealer and bookbinder, and at the age of 14 he was apprenticed to the man. Unlike the other apprentices, Faraday took the opportunity to read some of the books brought in for rebinding. The article on electricity in the third edition of the *Encyclopædia Britannica* particularly fascinated him.

Faraday's great opportunity came when he was offered a ticket to attend chemical lectures by Sir Humphry Davy at the Royal Institution of Great Britain in London. Faraday went, sat spellbound, recorded the lectures in his notes, and returned to bookbinding with the seemingly unrealizable hope of entering the temple of science. He sent a bound copy of his notes to Davy along with a letter asking for employment, but there was no opening. Davy did not forget, however, and, when one of his laboratory assistants was dismissed for brawling, he offered Faraday a job. Faraday began as Davy's laboratory assistant and learned chemistry at the elbow of one of the greatest practitioners of the day. It has been said, with some truth, that Faraday was Davy's greatest discovery.

When Faraday joined Davy in 1812, Davy was in the process of revolutionizing the chemistry of the day. Faraday's apprenticeship under Davy came to an end in 1820, but by then he had learned chemistry as thoroughly as anyone alive. He had also had ample opportunity to practice chemical analyses and laboratory techniques to the point of complete mastery, and he had developed his theoretical views to the point that they could guide him in his researches. There followed a series of discoveries that astonished the scientific world.

Faraday achieved his early renown as a chemist. His reputation as an analytical chemist led to his being called as an expert witness in legal trials and to the building up of a clientele whose fees helped to support the Royal Institution. In 1820 he produced the first known compounds of carbon and chlorine, C_2Cl_6 and C_2Cl_4. In 1825, as a result of research on illuminating gases, Faraday isolated and described benzene. In the 1820s he also conducted investigations of steel alloys, helping to lay the foundations for scientific metallurgy and metallography. While completing an assignment from the Royal Society to improve the quality of optical glass for telescopes, he produced a glass of very high refractive index that was to lead him, in 1845, to the discovery of diamagnetism.

In 1820 Hans Christian Ørsted had announced the discovery that the flow of an electric current through a wire produced a magnetic field around the wire. André-Marie Ampère then showed that the magnetic force apparently was a circular one, producing in effect a cylinder of magnetism around the wire. No such circular force had ever before been observed, and Faraday was the first to understand what it implied. If a magnetic pole could be isolated, it ought to move constantly in a circle around a current-carrying wire. Faraday's ingenuity and laboratory skill enabled him to construct an apparatus that confirmed this conclusion. The device, which transformed electrical energy into mechanical energy, was the first electric motor. This discovery led Faraday to contemplate the nature of electricity.

Unlike his contemporaries, he was not convinced that electricity was a material fluid that flowed through wires like water through a pipe. Instead, he thought of it as a vibration or force that was somehow transmitted as the result of tensions created in the conductor. Early experiments were unsuccessful until he began to work with Charles (later Sir Charles) Wheatstone on the theory of sound, another vibrational phenomenon. On August 29 1831 Faraday wound a thick iron ring on one side with insulated wire that was connected to a battery. He then wound the opposite side with wire connected to a galvanometer. What he expected was that a "wave" would be produced when the battery circuit was closed and that the wave would show up as a deflection of the galvanometer in the second circuit. He closed the primary circuit and, to his delight and satisfaction, saw the galvanometer needle jump. A current had been induced in the secondary coil by one in the primary.

Faraday then attempted to determine just how an induced current was produced. His original experiment had involved a powerful electromagnet, created by the winding of the primary coil. He now tried to create a current by using a permanent magnet. He discovered that when a permanent magnet was moved in and out of a coil of wire a current was induced in the coil. He immediately realized that a continuous current could be produced by rotating a copper disk between the poles of a powerful magnet and taking leads off the disk's rim and centre. The outside of the disk would cut more lines of magnetic force than would the inside, and there would thus be a continuous current produced in the circuit linking the rim to the centre. This was the first dynamo. It was also the direct ancestor of electric motors, for it was only necessary to reverse the situation, to feed an electric current to the disk, to make it rotate.

While Faraday was performing these experiments and presenting them to the scientific world, doubts were raised about the identity of the different manifestations of electricity that had been studied. Understanding voltaic and electromagnetic electricity posed no problems, but static electricity did. As Faraday delved deeper into the problem, he made some startling discoveries. These led him to a new theory of electrochemistry: (1) the amount of a substance deposited on each electrode of an electrolytic cell is directly proportional to the quantity of electricity passed through the cell, and (2) the quantities of different elements deposited by a given amount of electricity are in the ratio of their chemical equivalent weights.

By 1839 Faraday was able to bring forth a new and general theory of electrical action. Electricity, whatever it was, caused tensions to be created in matter. When these tensions were rapidly relieved (i.e. when bodies could not take much strain before "snapping" back), then what occurred was a rapid repetition of a cyclical build-up, breakdown, and build-up of tension that, like a wave, was passed along the substance. Such substances were called conductors. In electrochemical processes the rate of build-up and breakdown of the strain was proportional to the chemical affinities of the substances involved, but again the current was not a material flow but a wave pattern of tensions and their relief. Insulators were simply materials whose particles could take an extraordinary amount of strain before they snapped. Electrostatic charge in an isolated insulator was simply a measure of this accumulated strain. Thus, all electrical action was the result of forced strains in bodies.

The strain on Faraday of eight years of sustained experimental and theoretical work was too much, and in 1839 his health broke down. For the next six years he did little creative science, and not until 1845 was he able to pick up the thread of his researches and extend his theoretical views. When he returned to active research in 1845, it was to tackle again a problem that had obsessed him for years, that of his hypothetical electrotonic state. He was still convinced that it must exist and that he simply had not yet discovered the means for detecting it. Once again he tried to find signs of intermolecular strain in substances through which electrical lines of force passed, but again with no success. It was at this time that a young Scot, William Thomson (later Baron Kelvin), wrote to Faraday that he had studied Faraday's papers on electricity and magnetism and that he, too, was convinced that some kind of strain must exist. He suggested that Faraday experiment with magnetic lines of force, since these could be produced at much greater strengths than could electrostatic ones.

Faraday took the suggestion, passed a beam of plane-polarized light through the optical glass of high refractive index that he had developed in the 1820s, and then turned on an electromagnet so that its lines of force ran parallel to the light ray. This time he was rewarded with success. The plane of polarization was rotated, indicating a strain in the molecules of the glass. But Faraday again noted an unexpected result. When he changed the direction of the ray of light, the rotation remained in the same direction – a fact that Faraday correctly interpreted as meaning that the strain was not in the molecules of the glass but in the magnetic lines of force. The direction of rotation of the plane of polarization depended solely upon the polarity of the lines of force; the glass served merely to detect the effect.

This discovery confirmed Faraday's faith in the unity of forces, and he plunged onward, certain that all matter must exhibit some response to a magnetic field. To his surprise he found that this was in fact so, but in a peculiar way. Some substances, such as iron, nickel, cobalt, and oxygen, lined up in a magnetic field so that the long axes of their crystalline or molecular structures were parallel to the lines of force; others lined up perpendicular to the lines of force. Substances of the first class moved towards more intense magnetic fields; those of the second moved towards regions of less magnetic force. Faraday named the first group paramagnetics and the second diamagnetics. After further research he concluded that paramagnetics were bodies that conducted magnetic lines of force better than did the surrounding medium, whereas diamagnetics conducted them less well. By 1850 Faraday had evolved a radically new view of space and force. Space was not "nothing", the mere location of bodies and forces, but a medium capable of supporting the strains of electric and magnetic forces. The energies of the world were not localized in the particles from which these forces arose, but rather were to be found in the space surrounding them. Thus was born field theory. As James Clerk Maxwell (1831–79) later freely admitted, the basic ideas for his mathematical theory of electrical and magnetic fields came from Faraday; his contribution was to mathematize those ideas in the form of his classical field equations.

From about 1855 Faraday's mind began to fail. He still did occasional experiments, one of which involved attempting to find an electrical effect of raising a heavy weight – since he felt that gravity, like magnetism, must be convertible into some other force, most likely electrical. This time he was disappointed in his expectations, and the Royal Society refused to publish his negative results. Faraday began to sink into senility. Queen Victoria rewarded his lifetime of devotion to science by granting him the use of a house at Hampton Court and even offered him the honour of a knighthood. Faraday gratefully accepted the cottage but rejected the knighthood; he would, he said, remain plain Mr Faraday to the end.

CHARLES BABBAGE (1791–1871) AND ADA KING, COUNTESS OF LOVELACE (1815–1852)

English mathematician credited with having conceived the first automatic digital computer, and his associate who helped develop the prototype computer program.

Babbage was born in Teignmouth, Devon. At the age of 19 he helped found the Analytical Society, whose purpose was to introduce developments from Europe into English mathematics. In 1816 he was elected a fellow of the Royal Society, and he was soon instrumental in founding the Royal Astronomical (1820) and Statistical (1834) Societies. As a founding member of the Royal Astronomical Society, Babbage had seen a clear need to design and build a mechanical device that could automate long, tedious astronomical calculations. He began by writing a letter in 1822 to Sir Humphry Davy, President of the Royal Society, about the possibility of automating the construction of mathematical tables – specifically, logarithm tables for use in navigation. He then wrote a paper, "On the Theoretical Principles of the Machinery for Calculating Tables", which he read to the society later that year. (It won the Royal Society's first Gold Medal in 1823.) Tables then in use often contained errors, which could be a life-and-death matter for sailors at sea, and Babbage argued that by automating the production of the tables he could assure their accuracy. Having gained support in the society for his difference engine, as he called it, Babbage next turned to the British government to fund development, obtaining one of the world's first government grants for research and technological development.

Babbage approached the project very seriously: he hired a master machinist, set up a fireproof workshop, and built a dustproof environment for testing the device. Up until then calculations were rarely carried out to more than six digits, but Babbage planned to produce 20- or 30-digit results routinely. The difference engine was a digital device: it operated on discrete digits rather than continuously varying quantities, and the digits were decimal (0 to 9), represented by positions on toothed wheels. When one of the toothed wheels turned from 9 to 0, it caused the next wheel to advance one position.

The difference engine was more than a simple calculator, however. It mechanized not just a single calculation but a whole series of calculations on a number of variables to solve a complex problem, and went far beyond calculators in other ways too. Like modern computers, the difference engine had storage – that is, a place where data could be held temporarily for later processing – and it was designed to stamp its output into soft metal, which could later be used to produce a printing plate.

The full engine, designed to be room-sized, was never built, at least not by Babbage. By the time the funding for the difference engine had run out in 1833, however, he had conceived of something far more revolutionary: a general-purpose computing machine called the analytical engine. The analytical engine was to be a general-purpose, fully programme-controlled, automatic mechanical digital computer: it would be able to perform any calculation set before it. Before Babbage there is no evidence that anyone had ever conceived of such a device, let alone attempted to build one. The machine was designed to consist of four components: the mill, the store, the reader, and the printer – which are the essential components of every computer today. The mill was the calculating unit, analogous to the central processing unit (CPU) in a modern computer; the store was where data were held prior to processing, exactly analogous to memory and storage in today's computers; and the reader and printer were the input and output devices.

As with the difference engine, the proposed machine was far more complex than anything previously built. The store was to be large enough to hold 1,000 50-digit numbers – larger than the storage capacity of any computer built before 1960. The machine was to be steam driven and run by one attendant. The printing capability was also ambitious, and Babbage wanted to automate the process as much as possible, right up to producing printed tables of numbers. The reader was another new feature of the analytical engine. Data (numbers) were to be entered on punched cards, using the card-reading technology of the Jacquard loom. Instructions were also to be entered on cards, another idea taken directly from Jacquard.

By most definitions, the analytical engine was a real computer as understood today – or would have been, had not Babbage run into implementation problems again. Actually building his ambitious design was judged unfeasible given the current technology, and Babbage's failure to generate the promised mathematical tables with his difference engine had dampened enthusiasm for further government funding. Indeed, it was apparent to the British government that Babbage was more interested in innovation than in constructing tables.

All the same, Babbage's analytical engine was revolutionary. Its most radical feature was the ability to change its operation by changing the instructions on punched cards. Until this breakthrough, all the mechanical aids to calculation were merely calculators – or, like the difference engine, glorified calculators. The distinction between calculator and computer, although clear to Babbage, was not apparent to most people in the early nineteenth century, even to the intellectually adventuresome visitors at Babbage's soirées – with the exception of a young girl of unusual parentage and education.

Augusta Ada King, the Countess of Lovelace, was the daughter of the poet Lord George Gordon Byron and the mathematically inclined Anne Millbanke. One of her tutors was Augustus De Morgan, a famous mathematician and logician. Because Byron was involved in a notorious scandal at the time of her birth, Ada's mother encouraged her mathematical and scientific interests, hoping to suppress any inclination to wildness she may have inherited from her father.

Lady Lovelace attended Babbage's soirées and became fascinated with his difference engine. She also corresponded with him, asking pointed questions. It was his plan for the analytical engine that truly fired her imagination, however. In 1843, at the age of 27, she had come to understand it well enough to publish the definitive paper explaining the device and drawing the crucial distinction between this new thing and existing calculators. The analytical engine, she argued, went beyond the bounds of arithmetic. Because it operated on general symbols rather than on numbers, it established "a link . . . between the operations of matter and the abstract mental processes of the most abstract branch of mathematical science." It was a physical device that was capable of operating in the realm of abstract thought. Lady Lovelace rightly reported that this was not only something no one before had built; it was something that no one before had even conceived. She went on to become the world's only expert on the process of sequencing instructions on the punched cards that the analytical engine used – that is, she became the world's first computer programmer.

One feature of the analytical engine was its ability to place numbers and instructions temporarily in its store and return them to its mill for processing at an appropriate time. This was accomplished by the proper sequencing of instructions and data in its reader. Furthermore, the ability to reorder instructions and data gave the machine a flexibility and power that was hard to grasp – the first electronic digital computers of a century later lacked this ability. It was remarkable that a young scholar realized its importance in 1840, and it would be 100 years before anyone would understand it so well again. In the intervening century, attention would be diverted to the calculator and other business machines.

SIR CHARLES LYELL, BARONET (1797–1875)

Scottish geologist largely responsible for the general acceptance of the view that all features of the Earth's surface are produced over long periods of time.

Lyell was born at Kinnordy, the stately family home at the foot of the Grampian Mountains in eastern Scotland. At the age of 19 he entered Oxford University, where his interest in classics, mathematics, and geology was stimulated – the latter by the enthusiastic lectures of William Buckland, later widely known for his attempt to prove Noah's Flood by studies of fossils from cave deposits. Lyell spent the long student holidays travelling and conducting geological studies – notes made in 1817 on the origin of the Yarmouth lowlands clearly foreshadow his later work. The penetrating geological and cultural observations Lyell made while on a continental tour with his family in 1818 were as remarkable as the number of miles he walked in a day. In December 1819 he earned a BA with honours and moved to London to study law.

Lyell's eyes were weakened by hard study, and he sought and found relief by spending much time on geological work outdoors. In 1823, on a visit to Paris, he met the renowned naturalists Alexander von Humboldt and Georges Cuvier and examined the Paris Basin with the French geologist Louis-Constant Prévost. In 1824 Lyell studied sediments forming in freshwater lakes near Kinnordy.

Encouraged to finish his law studies, Lyell was admitted to the bar in 1825, but with his father's financial support he practiced geology more than law, publishing his first scientific papers that year. he was rapidly developing new principles of reasoning in geology and began to plan a book that would stress that there are natural (as opposed to supernatural) explanations for all geologic phenomena, that the ordinary natural processes of today and their products do not differ in kind or magnitude from those of the past, and that the Earth must therefore be very ancient because these everyday processes work so slowly.

The first volume of Lyell's *Principles of Geology* was published in July 1830. A reader today may wonder why this book filled with facts purports to deal with principles, but Lyell had to teach his principles through numerous facts and examples because in 1830 his method of scientific inquiry was novel and even mildly heretical. A remark of Charles Darwin (1809–82) shows how brilliantly Lyell succeeded: "The very first place which I examined . . . showed me clearly the wonderful superiority of Lyell's manner of treating geology, compared with that of any other author, whose work I had with me or ever afterwards read."

During the summer of 1830 Lyell travelled through the geologically complex Pyrenees to Spain. Back in London he set to work again on the *Principles of Geology*, finishing Volume II in December 1831 and the third and final volume in April 1833. During the next several years his winters were devoted to study; scientific and social activities; and revision of *Principles of Geology*, which sold so well that new editions were frequently required. Data for the new editions were gathered during summer travels, including two visits to Scandinavia in 1834 and 1837. In 1832 and 1833 Lyell delivered well-received lectures at King's College, London.

Publication of the *Principles of Geology* placed him among the recognized leaders of his field, compelling him to devote more time to scientific affairs. In 1838 Lyell's *Elements of Geology* was published: it described European rocks and fossils from the most recent, Lyell's speciality, to the oldest then known. Like the *Principles of Geology*, this well-illustrated work was periodically enlarged and updated.

In the 1840s Lyell became more widely known outside the scientific community while his professional reputation continued to grow; during his lifetime he received many awards and honorary degrees – including, in 1858, the Copley Medal, the highest award of the Royal Society, and he was many times president of various scientific societies or functions. However, his expanding reputation and responsibilities did not lessen his geological explorations.

In 1859 the publication of Darwin's *Origin of Species* gave new impetus to Lyell's work. Although Darwin drew heavily on Lyell's *Principles of Geology* both for style and content,

Lyell had never shared his protégé's belief in evolution. Why Lyell was hesitant in accepting Darwinism is best explained by Darwin himself: "Considering his age, his former views, and position in society, I think his action has been heroic."

LOUIS AGASSIZ (1807–1873)

Swiss-born US naturalist, geologist, and teacher who made revolutionary contributions to the study of glacier activity and extinct fishes.

Agassiz was the son of the Protestant pastor of Motier, a village on the shore of Lake Morat, Switzerland. He entered the universities of Zürich, Heidelberg, and Munich and took at Erlangen the degree of doctor of philosophy and at Munich that of doctor of medicine. His interest in ichthyology began with his study of an extensive collection of Brazilian fishes, mostly from the Amazon River, which had been collected in 1819 and 1820 by two eminent naturalists at Munich. The classification of these species was begun by one of the collectors in 1826, and when he died the collection was turned over to Agassiz. The work was completed and published in 1829 as *Selecta Genera et Species Piscium*. The study of fish forms became henceforth the prominent feature of Agassiz's research. In 1830 he issued a prospectus of a *History of the Fresh Water Fishes of Central Europe*, printed in parts from 1839 to 1842.

The year 1832 was significant in Agassiz's early career because it took him to Paris, where he lived the life of an impecunious student in the Latin Quarter, supporting himself and helped at times by the kindly interest of such friends as the

German naturalist Alexander von Humboldt – who secured for him a professorship at Neuchâtel, Switzerland – and Baron Cuvier, the most eminent ichthyologist of his time. Already Agassiz had become interested in the rich stores of the extinct fishes of Europe, especially those of Glarus in Switzerland and of Monte Bolca near Verona – of which, at that time, only a few had been critically studied. As early as 1829 he planned a comprehensive and critical study of these fossils and spent much time gathering material wherever possible.

Agassiz's epoch-making work, *Recherches sur les poissons fossiles* ("Researches on Fossil Fishes"), appeared in parts from 1833 to 1843. In it, the number of named fossil fishes was raised to more than 1,700, and the ancient seas were made to live again through the descriptions of their inhabitants. The great importance of this fundamental work rests on the impetus it gave to the study of extinct life itself. Turning his attention to other extinct animals found with the fishes, Agassiz published in 1838–42 two volumes on the fossil echinoderms of Switzerland, and later (1841–2) his *Études critiques sur les mollusques fossiles* ("Critical Studies on Fossil Mollusks").

In 1836 Agassiz began a new line of studies: the movements and effects of the glaciers of Switzerland. Several writers had expressed the opinion that these rivers of ice had once been much more extensive and that the erratic boulders scattered over the region and up to the summit of the Jura Mountains were carried by moving glaciers. On the ice of the Aar Glacier he built a hut, the "Hôtel des Neuchâtelois", from which he and his associates traced the structure and movements of the ice. In 1840 he published his *Études sur les glaciers* ("Studies on Glaciers"), in some respects his most important work. In it, Agassiz showed that at a geologically recent period Switzerland had been covered by one vast ice sheet. His final conclusion was that "great sheets of ice, resembling those now existing in Greenland, once covered all the countries in which unstratified gravel (boulder drift) is found."

In 1846 Agassiz visited the United States for the general purpose of studying natural history and geology there, but more specifically to give a course of lectures at the Lowell Institute in Boston. The lectures were followed by another series in Charleston and, later, by both popular and technical lectures in various cities. In 1847 he accepted a professorship of zoology at Harvard University.

In the United States his chief volumes of scientific research were the *Lake Superior* (1850); the *Contributions to the Natural History of the United States* (1857–62), in four volumes, the most notable being on the embryology of turtles; and the *Essay on Classification* (1859). Agassiz's industry and devotion to scientific pursuits continued, but two other traits now assumed importance. Quite possibly he was the ablest science teacher, administrator, promoter, and fund-raiser in the United States in the nineteenth century. In addition, he was devoted to his students, who were in the highest sense co-workers with him. Agassiz's method as a teacher was to provide contact with nature rather than information: he discouraged the use of books except in detailed research. The result of his instruction at Harvard was a complete revolution in the study of natural history in the United States.

In the interests of better teaching and of scientific enthusiasm, he organized in the summer of 1873 the Anderson School of Natural History at Penikese, an island in Buzzards Bay. This school, which had great influence on science teaching in America, was run solely by Agassiz and closed down after his death.

CHARLES DARWIN (1809–1882)

English naturalist whose theory of evolution
by natural selection became the foundation
of modern evolutionary studies.

Darwin was the second son of society doctor Robert Waring
Darwin and Susannah Wedgwood, daughter of the Unitarian
pottery industrialist Josiah Wedgwood. Darwin's paternal
grandfather, Erasmus Darwin, a freethinking physician and
poet fashionable before the French Revolution, was author of
Zoonomia or the Laws of Organic Life (1794–6). Darwin's
father sent him to study medicine at Edinburgh University in
1825. Later in life, Darwin gave the impression that he had
learned little during his two years at Edinburgh, but in fact it
was a formative experience. There was no better science
education in a British university, and radical students at the
university exposed Darwin to the latest Continental sciences.
The young Darwin learned much in Edinburgh's rich intellectual environment, but not medicine: he loathed anatomy, and
(pre-chloroform) surgery sickened him. His freethinking
father, shrewdly realizing that the church was a better calling
for an aimless naturalist, transferred him to Christ's College,
University of Cambridge, in 1828, where Darwin was educated as an Anglican gentleman and managed tenth place in
the bachelor of arts degree in 1831. Here he was shown the
conservative side of botany by a young professor, the Rever-
end John Stevens Henslow.

Fired by Alexander von Humboldt's account of the South
American jungles in his *Personal Narrative of Travels*, Darwin
jumped at Henslow's suggestion of a voyage to Tierra del
Fuego, at the southern tip of South America, aboard the rebuilt
brig HMS *Beagle*. Darwin would not sail as a lowly surgeon-

naturalist but as a self-financed gentleman companion to the 26-year-old captain, Robert Fitzroy, an aristocrat who feared the loneliness of command. Darwin equipped himself with weapons, books (Fitzroy gave him the first volume of *Principles of Geology*, by Charles Lyell [1797–1875]), and advice on preserving carcasses from London Zoo's experts. The *Beagle* sailed from England on 27 December 1831.

The circumnavigation of the globe would be the making of the 22-year-old Darwin. Five years of physical hardship and mental rigour, imprisoned within a ship's walls, offset by wide-open opportunities in the Brazilian jungles and the Andes Mountains, were to give Darwin a new seriousness. As a gentleman naturalist he could leave the ship for extended periods, pursuing his own interests. As a result, he spent only 18 months of the voyage aboard the ship.

On the Cape Verde Islands, in January 1832, Darwin saw bands of oyster shells running through local rocks, suggesting that Lyell was right in his geologic speculations and that the land was rising in places and falling in others. At Bahia (now Salvador), Brazil, the luxuriance of the rainforest left Darwin's mind in "a chaos of delight". Darwin always remembered with a shudder his observations of the parasitic ichneumon wasp, which stored caterpillars to be eaten alive by its grubs. He would later consider this to be evidence against the beneficent design of nature.

Darwin's periodic trips over two years to the cliffs at Bahía Blanca, Argentina, and further south at Port St Julian, yielded huge bones of extinct mammals. Darwin manhandled skulls, femurs, and armour plates back to the ship – relics, he assumed, of rhinoceroses, mastodons, cow-sized armadillos, and giant ground sloths. He unearthed a horse-sized mammal with a long face like an anteater's, and he returned from a 550-km (342-mile) ride to Mercedes near the Uruguay River with a skull 71 cm (28 inches) long strapped to his horse. Fossil extraction became a romance for Darwin. It pushed him into thinking of the primeval world and what had caused these giant beasts to die out.

Following an earthquake along the coast of Chile, Darwin was intrigued by the seemingly insignificant: the local mussel beds, all dead, were now lying above high tide. The continent was thrusting itself up, a metre or so at a time. He imagined the aeons it had taken to raise the fossilized trees in sandstone (once seashore mud) to 2,100 metres (6,900 feet) where he found them. Darwin began thinking in terms of deep time.

The *Beagle* left Peru on the circumnavigation home in September 1835. First Darwin landed on the "frying hot" Galapagos Islands. These were volcanic islands, crawling with marine iguanas and giant tortoises – but, contrary to legend, they never provided Darwin's "eureka" moment. Although he noted that the mockingbirds differed on four islands and tagged his specimens accordingly, he failed to label his other birds – what he thought were wrens, "gross-beaks", finches, and oriole-relatives – by island.

On the last leg of the voyage Darwin finished his 770-page diary, wrapped up 1,750 pages of notes, drew up 12 catalogues of his 5,436 skins, bones, and carcasses – and still he wondered: was each Galapagos mockingbird a naturally produced variety? Why did ground sloths become extinct? He sailed home with questions enough to last him a lifetime. After his return in October 1836, the supreme theorizer – who would always move from small causes to big outcomes – had the courage to look beyond the conventions of his own Victorian culture for new answers.

Darwin now settled down among the urban gentry as a gentleman geologist. He befriended Lyell, and he discussed the rising Chilean coastline as a new fellow of the Geological Society in January 1837 (he was secretary of the society by 1838). Darwin became well known through his diary's publication as *Journal of Researches into the Geology and Natural History of the Various Countries Visited by H.M.S. Beagle* (1839). With a £1,000 Treasury grant, obtained through the Cambridge network, he employed the best experts and published their descriptions of his specimens in his *Zoology of the Voyage of H.M.S. Beagle* (1838–43). Darwin's star had risen, and he was now lionized in London.

The experts' findings sent Darwin to more heretical depths. At the Royal College of Surgeons, the eminent anatomist Richard Owen found that Darwin's Uruguay River skull belonged to *Toxodon*, a hippotamus-sized antecedent of the South American capybara. The Pampas fossils were nothing like rhinoceroses and mastodons: they were huge extinct armadillos, anteaters, and sloths, which suggested that South American mammals had been replaced by their own kind according to some unknown "law of succession".

At the Zoological Society, the ornithologist John Gould announced that the Galapagos birds were not a mixture of wrens, finches, and "gross-beaks", but all ground finches, differently adapted. When Gould diagnosed the Galapagos mockingbirds as three species, unique to different islands, in March 1837, Darwin examined Fitzroy's collection to discover that each island had its representative finch as well. But how had they all diverged from mainland colonists?

By this time Darwin was living near his freethinking brother, Erasmus, in London's West End, and their dissident dining circle, which included the Unitarian Harriet Martineau, provided the perfect milieu for Darwin's ruminations. Darwin adopted "transmutation" (evolution, as it is now called), perhaps because of his familiarity with it through the work of his grandfather and the radical evolutionist Robert Grant. Nonetheless, it was abominated by the Cambridge clerics as a bestial, if not blasphemous, heresy that would corrupt mankind and destroy the spiritual safeguards of the social order. Thus began Darwin's double life, which would last for two decades.

For two years, with intensity and doggedness, he filled notebooks with jottings. He searched for the causes of extinction, accepted life as a branching tree (not a series of escalators, the old idea), tackled island isolation, and wondered whether variations appeared gradually or at a stroke. He became relativistic, sensing that life was spreading outward into niches, not standing on a ladder. There was no way of ranking humans and bees no yardstick of "highness": man was no longer the crown of creation. Darwin was also adopting ideas developed from Thomas Malthus's *Essay on the Principle of Population* (1838). Darwin called his modified Mathusian mechanism "natural selection" – positing that, when overpopulated, Nature experienced a fierce struggle, and from all manner of chance variation – good and bad – the best, "the surviving one of ten thousand trials," won out, endured, and thus passed on its improved trait.

Darwin was a born list maker. In 1838 he even totted up the pros and cons of taking a wife – and married his cousin Emma Wedgwood (1808–96) in 1839. He rashly confided his thoughts on evolution, evidently shocking her. Although the randomness and destructiveness of his evolutionary system, with thousands dying so that the "fittest" might survive, left little room for a personally operating benign deity, Darwin still believed that God was the ultimate lawgiver of the universe. In 1839 he shut his last major evolution notebook, his theory largely complete.

Darwin drafted a 35-page sketch of his theory of natural

selection in 1842 and expanded it in 1844, but he had no immediate intention of publishing it. He wrote Emma a letter in 1844 requesting that, if he died, she should pay an editor £400 to publish the work. Perhaps he wanted to die first. In 1842, Darwin, increasingly shunning society, had moved the family to the isolated village of Downe, in Kent, at the "extreme edge of [the] world". (It was in fact only 26 km [16 miles] from central London.) Here, living in a former parsonage, Down House, he emulated the lifestyle of his clerical friends.

He rarely mentioned his secret. When he did, notably to the Kew Gardens botanist Joseph Dalton Hooker, Darwin said that believing in evolution was "like confessing a murder". The analogy with this capital offence was not so strange: seditious atheists were using evolution as part of their weaponry against Anglican oppression and were being jailed for blasphemy. Darwin, nervous and nauseous, understood the conservative clerical morality.

From 1846 to 1854, Darwin added to his credibility as an expert on species by pursuing a detailed study of all known barnacles. Intrigued by their sexual differentiation, he discovered that some females had tiny degenerate males clinging to them. This sparked his interest in the evolution of diverging male and female forms from an original hermaphrodite creature. Four monographs on such an obscure group made him a world expert and gained him the Royal Society's Royal Medal in 1853. No longer could he be dismissed as a speculator on biological matters.

Through 1855 Darwin experimented with seeds in seawater, to prove that they could survive ocean crossings to start the process of speciation on islands. Then he kept fancy pigeons, to see if the chicks were more like the ancestral rock dove than their own bizarre parents. Darwin perfected his analogy of natural selection with the fancier's "artificial selection", as he called it. He was preparing his rhetorical strategy, ready to present his theory.

After speaking to Hooker and T.H. Huxley in Downe in April 1856, Darwin began writing a triple-volume book, tentatively called *Natural Selection*, which was designed to crush the opposition with a welter of facts. He had finished a quarter of a million words by 18 June 1858. That day he received a letter from Alfred Russel Wallace, an English socialist and specimen collector working in the Malay Archipelago, sketching a similar-looking theory. Darwin, fearing loss of priority, accepted Lyell's and Hooker's solution: they read joint extracts from Darwin's and Wallace's works at the Linnean Society on 1 July 1858. Darwin was away, sick, grieving for his tiny son who had died from scarlet fever, and thus he missed the first public presentation of the theory of natural selection. It was an absenteeism that would mark his later years.

Darwin hastily began an "abstract" of *Natural Selection*, which grew into a more accessible book, *On the Origin of Species by Means of Natural Selection, or the Preservation of Favoured Races in the Struggle for Life*. Suffering from a terrible bout of nausea, Darwin, now 50, was secreted away at a spa on the desolate Yorkshire moors when the book was sold to the trade on 22 November 1859. He still feared the worst and sent copies to the experts with self-effacing letters ("how you will long to crucify me alive"). It was like "living in Hell", he said about those months.

The newspapers drew the one conclusion that Darwin had specifically avoided: that humans had evolved from apes, and that Darwin was denying mankind's immortality. A sensitive Darwin, making no personal appearances, let Huxley (by now a good friend), manage this part of the debate. The pugnacious

Huxley, who loved public argument as much as Darwin loathed it, had his own reasons for taking up the cause – and did so with enthusiasm. He wrote three reviews of *Origin of Species* and defended human evolution.

Long periods of debilitating sickness in the 1860s left the craggy, bearded Darwin thin and ravaged. Down House was an infirmary where illness was the norm and Emma the attendant nurse. The house was also a laboratory, where Darwin continued experimenting and revamping the *Origin* through six editions. Although quietly swearing by "my deity 'Natural Selection'", he answered critics by re-emphasizing other causes of change. In *Variation of Animals and Plants under Domestication* (1868) he marshalled the facts and explored the causes of variation in domestic breeds by showing that fanciers picked from the gamut of naturally occurring variations to produce the tufts and topknots on their fancy pigeons.

Darwin was adept at flanking movements in order to get around his critics. He would take seemingly intractable subjects – such as orchid flowers – and make them test cases for natural selection. Hence the book that appeared after the *Origin* was, to everyone's surprise, *The Various Contrivances by which British and Foreign Orchids are Fertilised by Insects* (1862). He showed that the orchid's beauty was not a piece of floral whimsy "designed" by God to please humans but honed by selection to attract insect cross-pollinators. The petals guided the bees to the nectaries, and pollen sacs were deposited exactly where they could be removed by the stigma of another flower.

But why the importance of cross-pollination? Darwin's botanical work was always subtly related to his evolutionary mechanism. He believed that cross-pollinated plants would produce fitter offspring than self-pollinators, and he used considerable ingenuity in conducting thousands of crossings to prove the point. The results appeared in *The Effects of Cross and Self Fertilization in the Vegetable Kingdom* (1876). His next book, *The Different Forms of Flowers on Plants of the Same Species* (1877), was again the result of long-standing work into the way evolution in some species favoured different male and female forms of flowers to facilitate outbreeding. Darwin also studied insectivorous plants, climbing plants, and the response of plants to gravity and light (sunlight, he thought, activated something in the shoot tip, an idea that guided future work on growth hormones in plants).

Through the 1860s natural selection was already being applied to the growth of society: Wallace, for example, saw cooperation strengthening the moral bonds within primitive tribes. Advocates of social Darwinism, in contrast, complained that modern civilization was protecting the "unfit" from natural selection. Francis Galton, the anthropologist, argued that particular character traits – even drunkenness and genius – were inherited and that "eugenics", as it would come to be called, would stop the genetic drain. The trend to explain the evolution of human races, morality, and civilization was capped by Darwin in his two-volume *The Descent of Man, and Selection in Relation to Sex* (1871).

The book was authoritative, annotated, and heavily anecdotal in places. The first volume discussed human origins among the Old World monkeys and the growth of civilization. The second volume responded to critics like the eighth Duke of Argyll, who doubted that the iridescent hummingbird's plumage had any function – or any Darwinian explanation. Darwin argued that female birds were choosing mates for their gaudy plumage. As usual he tapped his huge correspondence network of breeders, naturalists, and travellers worldwide to produce evidence for this. Such "sexual selection" happened among humans too: with primitive societies accepting diverse notions of beauty, aesthetic preferences, he believed, could account for the origin of the human races.

Darwin finished another long-standing line of work. Since studying the moody orang-utans at London Zoo in 1838, and through the births of his ten children (whose facial contortions he duly noted), Darwin had been fascinated by expression. As a student he had heard the attacks on the idea that peoples' facial muscles were designed by God to express their unique thoughts. Now his photographically-illustrated *The Expression of the Emotions in Man and Animals* (1872) expanded the subject to include the rages and grimaces of asylum inmates, showing the continuity of emotions and expressions between humans and animals.

Darwin wrote his autobiography between 1876 and 1881. It was composed for his grandchildren, rather than for publication, and was particularly candid on his dislike of Christian myths of eternal torment. To people who inquired about his religious beliefs, however, he would only say that he was an agnostic (a word coined by Huxley in 1869).

The treadmill of experiment and writing gave so much meaning to Darwin's life. But, as he wrapped up his final, long-term interest, publishing *The Formation of Vegetable Mould, Through the Action of Worms* (1881), the future looked bleak. Such an earthy subject was typical Darwin: just as he had shown that today's ecosystems were built by infinitesimal degrees and the mighty Andes by tiny uplifts, so he ended on the monumental transformation of landscapes by the Earth's humblest denizens. Suffering from angina, he looked forward to joining the worms, contemplating "Down graveyard as the sweetest place on earth." He had a seizure in March 1882 and died of a heart attack on 19 April. Influential groups wanted a grander commemoration than a funeral in Downe, and Galton had the Royal Society request the family's permission for a state burial. Huxley convinced the canon of Westminster Abbey to bury the diffident agnostic there. And so Darwin was laid to rest with full ecclesiastical pomp on 26 April 1882, attended by the new nobility of science and the state.

ÉVARISTE GALOIS (1811–1832)

French mathematician noted for
his contributions to group theory.

Galois was the son of Nicolas-Gabriel Galois, an important citizen in the Paris suburb of Bourg-la-Reine. Galois was educated at home until 1823, when he entered the Collège Royal de Louis-le-Grand. There his education languished at the hands of mediocre and uninspiring teachers. But his mathematical ability blossomed when he began to study the works of his countrymen Adrien-Marie Legendre (1752–1833) on geometry and Joseph-Louis Lagrange (1736–1813) on algebra.

Under the guidance of Louis Richard, one of his teachers at Louis-le-Grand, Galois's further study of algebra led him to take up the question of the solution of algebraic equations. Mathematicians for a long time had used explicit formulae, involving only rational operations and extractions of roots, for the solution of equations up to the fourth degree, but they had been defeated by equations of the fifth degree and higher. In 1770 Lagrange took the novel but decisive step of treating the roots of an equation as objects in their own right and studying permutations (a change in an ordered arrangement) of them. In 1799 the Italian mathematician Paolo Ruffini attempted to prove the impossibility of solving the general quintic equation by radicals. Ruffini's effort was not wholly successful, but in 1824 the Norwegian mathematician Niels Abel gave a correct proof.

Galois, stimulated by Lagrange's ideas and initially unaware of Abel's work, began searching for the necessary and sufficient conditions under which an algebraic equation of any degree can be solved by radicals. His method was to analyse the "admissible" permutations of the roots of the equation. His key discovery, brilliant and highly imaginative, was that solvability by radicals is possible if and only if the group of automorphisms (functions that take elements of a set to other elements of the set while preserving algebraic operations) is solvable – which means essentially that the group can be broken down into simple "prime-order" constituents that always have an easily understood structure. The term "solvable" is used because of this connection with solvability by radicals. Thus, Galois perceived that solving equations of the quintic and beyond required a wholly different kind of treatment from that required for quadratic, cubic, and quartic equations. Although Galois used the concept of group and other associated concepts, such as coset and subgroup, he did not actually define these concepts, and he did not construct a rigorous formal theory.

While still at Louis-le-Grand Galois published one minor paper, but his life was soon overtaken by disappointment and tragedy. A memoir on the solvability of algebraic equations that he had submitted in 1829 to the French Academy of Sciences was lost by Augustin-Louis Cauchy. Galois failed in two attempts (1827 and 1829) to gain admission to the École Polytechnique, the leading school of French mathematics – his second attempt marred by a disastrous encounter with an oral examiner. Also in 1829 his father, after bitter clashes with conservative elements in his home town, committed suicide. The same year, Galois enrolled as a student teacher in the less prestigious École Normale Supérieure and turned to political activism. Meanwhile he continued his research, and in the spring of 1830 he had three short articles published. At the same time, he rewrote the paper that had been lost and presented it again to the Academy – but for a second time the manuscript went astray. Jean-Baptiste-Joseph Fourier took it home but died a few weeks later, and the manuscript was never found.

The July Revolution of 1830 sent the last Bourbon monarch, Charles X, into exile. But Republicans were deeply disappointed when yet another king, Louis-Philippe, ascended the throne – even though he was the "Citizen King" and wore the tricoloured flag of the French Revolution. When Galois wrote a vigorous article expressing pro-Republican views, he was promptly expelled from the École Normale Supérieure. Subsequently, he was arrested twice for Republican activities: he was acquitted the first time but spent six months in prison on the second charge.

In 1831 he presented his memoir on the theory of equations for the third time to the Academy. This time it was returned but with a negative report. The judges, who included Siméon-Denis Poisson, did not understand what Galois had written and (incorrectly) believed that it contained a significant error. They had been quite unable to accept Galois's original ideas and revolutionary mathematical methods.

The circumstances that led to Galois's death in a duel in Paris are not altogether clear, but recent scholarship suggests that it was at his own insistence that the duel was staged and fought to look like a police ambush. Whatever the case, in anticipation of his death the following day, the night before the duel Galois hastily wrote a scientific last testament addressed to his friend Auguste Chevalier, in which he summarized his work and included some new theorems and conjectures.

Galois's manuscripts, with annotations by Joseph Liouville, were published in 1846 in the *Journal de Mathématiques Pures et Appliquées* ("Journal of Pure and Applied Mathematics"). But it was not until 1870, with the publication of Camille Jordan's *Traité des Substitutions* ("Treatise on Substitutions"), that group theory became a fully established part of mathematics.

SIR FRANCIS GALTON (1822–1911)

English explorer, anthropologist, and eugenicist, known for his pioneering studies of human intelligence.

Galton was born near Sparkbrook in Birmingham, England. His parents had planned that he should study medicine, and a tour of medical institutions on the Continent in his teens – an unusual experience for a student of his age – was followed by training in hospitals in Birmingham and London. He attended Trinity College, University of Cambridge, but left without taking a degree and later continued his medical studies in London. But before they were completed, his father died, leaving him "a sufficient fortune to make me independent of the medical profession." Galton was then free to indulge his craving for travel. Leisurely expeditions in 1845–6 up the Nile with friends and into the Holy Land alone were preliminaries to a carefully organized penetration into unexplored parts of south-western Africa. After consulting the Royal Geographical

Society, Galton decided to investigate a possible opening from the south and west to Lake Ngami, which lies north of the Kalahari desert some 885 km (550 miles) east of Walvis Bay. The expedition, which included two journeys – one northward, the other eastward – from the same base, proved to be difficult and not without danger. Though the explorers did not reach Lake Ngami, they gained valuable information.

As a result, at the age of only 31, Galton was in 1853 elected a fellow of the Royal Geographical Society and, three years later, of the Royal Society. Galton wrote nine books and some 200 papers. They deal with many diverse subjects, including the use of fingerprints for personal identification; the correlational calculus, a branch of applied statistics (in both of which Galton was a pioneer); twins; blood transfusions; criminality; the art of travel in undeveloped countries; and meteorology. Most of Galton's publications disclose his predilection for quantifying: an early paper, for example, dealt with a statistical test of the efficacy of prayer. Moreover, over a period of 34 years, he concerned himself with improving standards of measurement.

Although he made contributions to many fields of knowledge, eugenics remained Galton's fundamental interest, and he devoted the latter part of his life chiefly to propagating the idea of improving the physical and mental makeup of the human species by selective parenthood. Galton, a cousin of Charles Darwin, was among the first to recognize the implications for mankind of Darwin's theory of evolution. He saw that it invalidated much of contemporary theology and that it also opened possibilities for planned human betterment. Galton coined the word "eugenics" to denote scientific endeavours to increase the proportion of persons with better-than-average genetic endowment through the careful selection of marriage partners.

In his *Hereditary Genius* (1869), in which he used the word "genius" to denote "an ability that was exceptionally high and at the same time inborn," Galton's main argument was that mental and physical features are equally inherited – a proposition that was not accepted at the time. It is surprising that when Darwin first read this book, he wrote to the author: "You have made a convert of an opponent in one sense for I have always maintained that, excepting fools, men did not differ much in intellect, only in zeal and hard work." This book doubtless helped Darwin to extend his evolution theory to humans.

Galton's *Inquiries into Human Faculty* (1883) consists of some 40 articles varying in length from 2 to 30 pages, which are mostly based on scientific papers written between 1869 and 1883. The book can in a sense be regarded as a summary of the author's views on the faculties of human beings. On all his topics, Galton has something original and interesting to say, and he says it with clarity, brevity, distinction, and modesty. Under the terms of his will, a eugenics chair was established at the University of London.

In the twentieth century Galton's name has been mainly associated with eugenics. Insofar as this field takes primary account of *inborn* differences between human beings, it has come under suspicion from those who hold that cultural (social and educational) factors heavily outweigh inborn, or biological, factors in their contribution to human differences. Eugenics is accordingly often treated as an expression of class prejudice, and Galton as a reactionary. Yet to some extent this view misrepresents his thought, for his aim was not the creation of an aristocratic elite but of a population consisting entirely of superior men and women. His ideas, like those of Darwin, were limited by a lack of an adequate theory of inheritance; the rediscovery of the work of Mendel came too late to affect Galton's contribution in any significant way.

GREGOR MENDEL (1822–84)

Austrian botanist, teacher, and Augustinian prelate;
the first to lay the mathematical foundation
of the science of genetics.

Mendel was born to a family with limited means in Heinzendorf, Austria (now Hyncice, Czech Rep.). His academic abilities were recognized by the local priest, who persuaded his parents to send Mendel away to school at the age of 11. As his father's only son, Mendel was expected to take over the small family farm, but Mendel chose to enter the Altbrünn Monastery as a novitiate of the Augustinian order, where he was given the name Gregor. The move to the monastery took Mendel to Brünn, the capital of Moravia, where for the first time he was freed from the harsh struggle of former years. He was also introduced to a diverse and intellectual community. As a priest, Mendel found his parish duty to visit the sick and dying so distressing that he became ill. Abbot Cyril Napp found him an alternative vocation: a teaching position at Znaim (now Znojmo, Czech Republic), where he proved very successful. However, in 1850, Mendel failed an exam, introduced through new legislation for teacher certification, and was sent to the University of Vienna for two years to benefit from a new programme of scientific instruction.

Mendel devoted his time at Vienna to physics and mathematics, working under the Austrian physicist Christian Doppler and mathematical physicist Andreas von Ettinghausen. He also studied the anatomy and physiology of plants and the use of the microscope under botanist Franz Unger, an enthusiast for the cell theory and a supporter of the developmentalist (pre-Darwinian) view of the evolution of life. In the summer of 1853, Mendel returned to the monastery in Brünn, and in the following year he was again given a teaching position, this time at the Brünn Realschule (secondary school), where he remained until he was elected Abbot 14 years later.

In 1854, Abbot Cyril Napp permitted Mendel to plan a major experimental programme in hybridization at the monastery. The aim of the programme was to trace the transmission of hereditary characters in successive generations of hybrid progeny. Previous authorities had observed that progeny of fertile hybrids tended to revert to the originating species, and they had therefore concluded that hybridization could not be a mechanism used by nature to multiply species – although in exceptional cases some fertile hybrids did appear not to revert (the so-called "constant hybrids"). On the other hand, plant and animal breeders had long shown that cross-breeding could indeed produce a multitude of new forms. The latter point was of particular interest to landowners, including the abbot of the monastery, who was concerned about the monastery's future profits from the wool of its Merino sheep, owing to competing wool being supplied from Australia.

Mendel chose to conduct his studies with the edible pea (*Pisum sativum*) because of the numerous distinct varieties, the ease of culture and control of pollination, and the high proportion of successful seed germinations. From 1854 to 1856 he tested 34 varieties for constancy of their traits. In order to trace the transmission of characters, he chose seven traits that were expressed in a distinctive manner, such as plant height (short or tall) and seed colour (green or yellow). He referred to these alternatives as contrasted characters, or character-pairs.

Mendel crossed varieties that differed in one trait – for instance, tall crossed with short. The first generation of hybrids (F_1) displayed the character of one variety but not that of the other. In Mendel's terms, one character was dominant and the other recessive. In the numerous progeny that he raised from these hybrids (the second generation, F_2), however, the recessive character reappeared, and the proportion of offspring bearing the dominant character to those bearing the recessive character was very close to a 3:1 ratio. Study of the descendants (F_3) of the dominant group showed that one-third of them were true-breeding and two-thirds were of hybrid constitution. The 3:1 ratio could hence be rewritten as 1:2:1, meaning that 50 per cent of the F_2 generation were true-breeding and 50 percent were still hybrid.

This was Mendel's major discovery. It was unlikely to have been made by his predecessors, since they did not grow statistically significant populations; nor did they follow the individual characters separately to establish their statistical relations. Mendel realized further that he could test his expectation that the seven traits are transmitted independently of one another. Crosses involving first two and then three of his seven traits yielded categories of offspring in proportions following the terms produced from combining two binomial equations, indicating that their transmission was independent of one another. Mendel's successors have called this conclusion the law of independent assortment.

Mendel went on to relate his results to the cell theory of fertilization, according to which a new organism is generated from the fusion of two cells. In order for pure breeding forms of both the dominant and the recessive type to be brought into the hybrid, there had to be some temporary accommodation of the two differing characters in the hybrid as well as a separation process in the formation of the sperm cells and egg cells. In other words, the hybrid must form germ cells bearing the potential to yield either the one characteristic or the other. This has since been described as the law of segregation, or the doctrine of the purity of the germ cells.

Mendel first presented his results in two separate lectures in 1865 to the Natural Science Society in Brünn. His paper "Experiments on Plant Hybrids" *was published the following year in the society's journal,* Verhandlungen des naturforschenden Vereines in Brünn. *It attracted little attention, although many libraries received it and reprints were sent out. The tendency of those who read it was to conclude that Mendel had simply demonstrated more accurately what was already widely assumed – namely, that hybrid progeny revert to their originating forms. They overlooked the potential for variability and the evolutionary implications that his demonstration of the recombination of traits made possible. Mendel appears to have made no effort to publicize his work, and by 1871 he had only enough time to continue his meteorological and apicultural (beekeeping) work.*

LOUIS PASTEUR (1822–1895)

French chemist and microbiologist who made valuable contributions to science and to industry, including the process known as pasteurization.

Pasteur was born in Dôle, France. He received a doctor of science degree in 1947 from the École Normale Supérieure, a noted teacher-training college in Paris, and became a professor of chemistry at the University of Strasbourg. In 1848 Pasteur presented before the Paris Academy of Sciences a paper reporting a remarkable discovery he had just made – that certain chemical compounds were capable of splitting into a "right" component and a "left" component, one component being the mirror image of the other. His discoveries arose out of a crystallographic investigation of tartaric acid, an acid formed in grape fermentation that is widely used commercially, and racemic acid – a new, hitherto unknown acid that had been discovered in certain industrial processes in the Alsace region. Both acids not only had identical chemical compositions but also had the same structure, yet they showed marked differences in properties. On the basis of these experiments, Pasteur elaborated his theory of molecular asymmetry, showing that the biological properties of chemical substances depend not only on the nature of the atoms constituting their molecules but also on the manner in which these atoms are arranged in space.

In 1854 Pasteur became dean of the new science faculty at the University of Lille, where he initiated a highly modern educational concept: by instituting evening classes for the many young workmen of the industrial city, conducting his regular students around large factories in the area, and organizing supervised practical courses, he demonstrated the relationship that he believed should exist between theory and practice; between university and industry. At Lille, after receiving a query from an industrialist on the production of alcohol from grain and beet sugar, Pasteur began his studies on fermentation. During his analysis he once again encountered – though in liquid form – new "right" and "left" compounds. By studying the fermentation of alcohol he went on to the problem of lactic fermentation, showing yeast to be an organism capable of reproducing itself, even in artificial media, without free oxygen – a concept that became known as the Pasteur effect.

In 1857 he was named Director of Scientific Studies at the École Normale Supérieure. He continued his researches and announced that fermentation was the result of the activity of minute organisms and that when fermentation failed, either the necessary organism was absent or was unable to grow properly. Before this discovery, all explanations of fermentation had lacked experimental foundation. Pasteur showed that milk could be soured by injecting a number of organisms from buttermilk or beer but could be kept unchanged if such organisms were excluded.

He was elected to the Academy of Sciences in 1862, and the following year a chair at the École des Beaux-Arts was established for him for a new and original programme of instruction in geology, physics, and chemistry applied to the fine arts. As a logical sequel to his work on fermentation, Pasteur began research on spontaneous generation (the concept that bacterial life arose spontaneously) – a question that at that time divided scientists into two opposing camps. Pasteur's recognition of the fact that both lactic and alcohol fermentations were hastened by exposure to air led him to wonder whether his invisible organisms were always present in the atmosphere or whether they were spontaneously generated. By means of simple and precise experiments, including the filtration of air and the exposure of unfermented liquids to the air of the high Alps, he proved that food decomposes when placed in contact with germs present in the air, which cause its putrefaction, and that it does not undergo transformation or putrefy in such a way as to spontaneously generate new organisms within itself.

After laying the theoretical groundwork, Pasteur proceeded to apply his findings to the study of vinegar and wine: two commodities of great importance in the economy of France. His pasteurization process – the destruction of harmful germs by heat – made it possible to produce, preserve, and transport these products without their undergoing deterioration.

Although Pasteur was partially paralyzed in 1868 and

applied for retirement from the university, he continued his research. In 1870 he devoted himself to the problem of beer spoilage. Following an investigation conducted both in France and among the brewers in London, he devised, as he had done for vinegar and wine, a procedure for manufacturing beer that would prevent its deterioration with time. British exporters, whose ships had to sail entirely around the African continent, were thus able to send British beer as far as India without fear of its deteriorating. In 1873 Pasteur was elected a member of the Academy of Medicine, and in 1874 the French Parliament provided him with an award that would ensure his material security while he pursued his work.

When in 1881 he had perfected a technique for reducing the virulence of various disease-producing microorganisms, he succeeded in vaccinating a herd of sheep against the disease known as anthrax. Likewise, he was able to protect fowl from chicken cholera, for he had observed that once animals stricken with certain diseases had recovered they were later immune to a fresh attack.

On 27 April 1882, Pasteur was elected a member of the Académie Française, at which point he undertook research that proved to be the most spectacular of all – the preventive treatment of rabies. After experimenting with inoculations of saliva from infected animals, he came to the conclusion that the virus was also present in the nerve centres, and he demonstrated that a portion of the *medulla oblongata* (lower brain stem) of a rabid dog, when injected into the body of a healthy animal, produced symptoms of rabies. By further work on the dried tissues of infected animals and the effect of time and temperature on these tissues, he was able to obtain a weakened form of the virus that could be used for inoculation.

Having detected the rabies virus by its effects on the nervous system and attenuated its virulence, he applied his procedure to humans: on 6 July 1885 he saved the life of a nine-year-old boy, Joseph Meister, who had been bitten by a rabid dog. The experiment was an outstanding success, opening the road to protection from a terrible disease. In 1888 the Pasteur Institute was inaugurated in Paris for the purpose of undertaking fundamental research, prevention, and treatment of rabies. Pasteur, although in failing health, headed the institute until his death in 1895.

ALFRED RUSSEL WALLACE (1823–1913)

British naturalist, geographer, and social critic whose formulation of the theory of evolution by natural selection predated Charles Darwin's published contributions.

Wallace grew up in modest circumstances in rural Wales and then in Hertford, England. His formal education was limited to six years at the one-room Hertford Grammar School. Although his education was curtailed by the family's worsening financial situation, his home was a rich source of books, maps, and gardening activities, which Wallace remembered as enduring sources of learning and pleasure.

In 1837 Wallace became an apprentice in the surveying business of his eldest brother, William. For approximately eight of the next ten years, he surveyed and mapped in Bedfordshire and then in Wales. He lived among farmers and artisans and saw the injustices suffered by the poor as a result of the recently instituted tax laws. Wallace's detailed observations of the locals' habits are recorded in one of his first writing efforts, an essay on "the South Wales Farmer".

Wallace spent a great deal of time outdoors, both in his surveying work and for pleasure. An enthusiastic amateur naturalist with an intellectual bent, he read widely in natural history, history, and political economy, including works by William Swainson, Charles Darwin, Alexander von Humboldt, and Thomas Malthus. He also read works and attended lectures on phrenology (the inference of personality traits from the shape of the skull's surface) and mesmerism (hypnosis), forming an interest in nonmaterial mental phenomena that grew increasingly prominent later in his life. Inspired by reading about organic evolution in Robert Chambers's controversial *Vestiges of the Natural History of Creation* (1844), unemployed, and ardent in his love of nature, Wallace and his naturalist friend Henry Walter Bates, who had introduced Wallace to entomology four years earlier, travelled to Brazil in 1848 as self-employed specimen collectors.

Wallace and Bates participated in the culture of natural history collecting – honing practical skills to identify, collect, and send back to England biological objects that were highly valued in the flourishing trade in natural specimens. The two young men amicably parted ways after several joint collecting ventures. Except for one shipment of specimens sent to his agent in London, however, most of Wallace's collections were lost on his voyage home when his ship went up in flames and sank. Nevertheless, he managed to save some of his notes before his rescue and return journey. From these he published several scientific articles, two books (*Palm Trees of the Amazon and Their Uses* and *Narrative of Travels on the Amazon and Rio Negro*, both 1853), and a map depicting the course of the Negro River. These won him acclaim from the Royal Geographical Society, which helped to fund his next collecting venture, in the Malay Archipelago.

Wallace spent eight years in the Malay Archipelago, from 1854 to 1862, travelling among the islands, collecting biological specimens for his own research and for sale, and writing scores of scientific articles on mostly zoological subjects. Among these were two extraordinary articles dealing with the origin of new species. The first of these, published in 1855, concluded with the assertion that "every species has come into existence coincident both in space and time with a pre-existing closely allied species." Wallace then proposed that new species arise by the progression and continued divergence of varieties that outlive the parent species in the struggle for existence.

In early 1858 he sent a paper outlining these ideas to Darwin, who saw such a striking coincidence to his own theory that he consulted his closest colleagues, the geologist Charles Lyell and the botanist Joseph Dalton Hooker. The three men decided to present two extracts of Darwin's previous writings, along with Wallace's paper, to the Linnean Society. The resulting set of papers, with both Darwin's and Wallace's names, was published as a single article entitled "On the Tendency of Species to Form Varieties; and on the Perpetuation of Varieties and Species by Natural Means of Selection" in the *Proceedings of the Linnean Society* in 1858. This compromise sought to avoid a conflict of priority interests and was reached without Wallace's knowledge. Wallace's research on the geographic distribution of animals among the islands of the Malay Archipelago provided crucial evidence for his evolutionary theories and led him to devise what soon became known as Wallace's Line, the boundary that separates the fauna of Australia from that of Asia.

Wallace returned to England in 1862 an established natural scientist and geographer, as well as a collector of more than 125,000 animal specimens. He published a highly successful narrative of his journey, *The Malay Archipelago: The Land of the Orang-Utan, and the Bird of Paradise* (1869), and wrote *Contributions to the Theory of Natural Selection* (1870). In the latter volume and in several articles from this period on human evolution and spiritualism, Wallace parted from the scientific naturalism of many of his friends and colleagues in claiming that natural selection could not account for the higher faculties of human beings.

Wallace's two-volume *Geographical Distribution of Animals* (1876) and *Island Life* (1880) became the standard authorities in zoogeography and island biogeography, synthesizing knowledge about the distribution and dispersal of living and extinct animals in an evolutionary framework.

Wallace received several awards, including the Royal Society's Royal Medal (1868), Darwin Medal (1890, for his independent origination of the theory of the origin of species by natural selection), Copley Medal (1908), and Order of Merit (1908); the Linnean Society's Gold Medal (1892) and Darwin-Wallace Medal (1908); and the Royal Geographical Society's Founder's Medal (1892). He was also awarded honorary doctorates from the Universities of Dublin (1882) and Oxford (1889) and won election to the Royal Society (1893). In 1881 he was added to the Civil List, thanks largely to the efforts of Darwin and T.H. Huxley.

WILLIAM THOMSON, BARON KELVIN (1824–1907)

Scottish engineer, mathematician, and physicist, who profoundly influenced the scientific thought of his generation.

Thomson was born in Belfast, Ireland. His father, who was a textbook writer, taught him the most recent mathematics, which had not yet become a part of the British university curriculum. At the age of ten William matriculated at the University of Glasgow, where he was introduced to the advanced and controversial mathematical work of Joseph Fourier (1768–1830). Thomson's first two published articles, which appeared when he was 16 and 17 years old, were a defence of Fourier's work, which was then under attack by British scientists. Thomson was the first to promote the idea that Fourier's mathematics, although applied solely to the flow of heat, could be used in the study of other forms of energy – whether fluids in motion or electricity flowing through a wire.

Thomson entered the University of Cambridge in 1841 and took his BA degree four years later with high honours. He then went to Paris, where he worked in the laboratory of the physicist and chemist Henri-Victor Regnault to gain practical, experimental competence to supplement his theoretical education. Thomson was appointed to the chair of natural philosophy (later called physics) at the University of Glasgow when the position fell vacant in 1846, and he remained at Glasgow for the rest of his career.

Thomson's scientific work was guided by the conviction that the various theories dealing with matter and energy were converging toward one great, unified theory. He pursued the goal of a unified theory even though he doubted that it was attainable in his lifetime or ever. The basis for his conviction was the cumulative impression obtained from experiments showing the interrelation of forms of energy. By the middle of the nineteenth century it had been shown that magnetism and electricity, electromagnetism, and light were related, and Thomson had shown by mathematical analogy that there was a relationship between hydrodynamic phenomena and an electric current flowing through wires. James Prescott Joule (1818–89) also claimed that there was a relationship between mechanical motion and heat, and his idea became the basis for the science of thermodynamics.

In 1847 Thomson first heard Joule's theory about the interconvertibility of heat and motion at a meeting of the British Association for the Advancement of Science. The theory went counter to the accepted knowledge of the time, which was that heat was an imponderable substance (caloric) and could not be, as Joule claimed, a form of motion. At the time, although he could not accept Joule's idea, Thomson was willing to reserve judgment – especially since the relation between heat and mechanical motion fit into his own view of the causes of force – and he was open-minded enough to discuss with Joule the implications of the new theory. By 1851 Thomson was able to give public recognition to Joule's theory, along with a cautious endorsement in a major mathematical treatise, *On the Dynamical Theory of Heat*. Thomson's essay contained his version of the second law of thermodynamics, which was a major step toward the unification of scientific theories.

Thomson's contributions to nineteenth-century science were many. He advanced the ideas of Michael Faraday (1791–1867), Fourier, Joule, and others. Using mathematical analysis, he drew generalizations from experimental results. He formulated the concept that was to be generalized into the dynamic theory of energy. He also advanced the frontiers of science in several other areas, particularly hydrodynamics; originated the mathematical analogy between the flow of heat in solid bodies and the flow of electricity in conductors; and developed the absolute temperature scale that became known as the Kelvin temperature scale.

Thomson's involvement in a controversy over the feasibility of laying a transatlantic cable changed the course of his professional work. His work on the project began in 1854 when Stokes, a lifelong correspondent on scientific matters, asked for a theoretical explanation of the apparent delay in an electric current passing through a long cable. In his reply, Thomson referred to his early paper "On the Uniform Motion of Heat in Homogeneous Solid Bodies, and its Connexion with the Mathematical Theory of Electricity" (1842). Thomson's idea about the mathematical analogy between heat flow and electric current worked well in his analysis of the problem of sending telegraph messages through the planned 4,800-km (3,000-mile) cable. His equations describing the flow of heat through a solid wire proved applicable to questions about the velocity of a current in a cable.

The publication of Thomson's reply to Stokes prompted a rebuttal by EOW Whitehouse, the Atlantic Telegraph Company's chief electrician. Whitehouse claimed that practical experience refuted Thomson's theoretical findings, and for a time Whitehouse's view prevailed with the directors of the company. Despite their disagreement, Thomson participated, as chief consultant, in the hazardous early cable-laying expeditions. In 1858 Thomson patented his telegraph receiver, called a mirror galvanometer, for use on the Atlantic cable. (The device, along with his later modification called the siphon recorder, came to be used on most of the worldwide network of submarine cables.) Eventually the directors of the Atlantic Telegraph Company fired Whitehouse, adopted Thomson's suggestions for the design of the cable, and decided in favour of the mirror galvanometer. Thomson was knighted in 1866 by Queen Victoria for his work.

After the successful laying of the transatlantic cable, Thomson became a partner in two engineering consulting firms, which played a major role in the planning and construction of submarine cables during the frenzied era of expansion that resulted in a global network of telegraph communication. He became a wealthy man who could afford a 126-tonne yacht and a baronial estate.

Thomson's interests in science included not only electricity, magnetism, thermodynamics, and hydrodynamics but also geophysical questions about tides, the shape of the Earth, atmospheric electricity, thermal studies of the ground, the Earth's rotation, and geomagnetism. He also entered the controversy over Charles Darwin's theory of evolution. Thomson challenged the views on geologic and biological change of the early Uniformitarians, including Darwin, who claimed that the Earth and its life had evolved over an incalculable number of years, during which the forces of nature always operated as at present. On the basis of thermodynamic theory and Fourier's studies, Thomson estimated in 1862 that more than one million years ago the sun's heat and the temperature of the Earth must have been considerably greater and that these conditions had produced violent storms and floods and an entirely different type of vegetation. Thomson's speculations as to the age of the Earth and the sun were inaccurate, but he did succeed in pressing his contention that biological and geologic theory had to conform to the well-established theories of physics.

Thomson's interest in the sea, roused aboard his yacht the *Lalla Rookh*, resulted in a number of patents: a compass that was adopted by the British Admiralty; a form of analogue computer for measuring tides in a harbour and for calculating tide tables for any hour, past or future; and sounding equipment. He established a company to manufacture these items and a number of electrical measuring devices. Like his father, he published a textbook, *Treatise on Natural Philosophy* (1867), a work on physics co-authored with Peter Guthrie Tait that helped shape the thinking of a generation of physicists.

Thomson was said to be entitled to more letters after his name than any man in the Commonwealth. He received honorary degrees from universities throughout the world and was lauded by engineering societies and scientific organizations. Elected a fellow of the Royal Society in 1851, he served as its president from 1890 to 1895. He published more than 600 papers and was granted dozens of patents.

JAMES CLERK MAXWELL (1831–1879)

Scottish physicist best known for his formulation of electromagnetic theory.

James Clerk Maxwell was born in Edinburgh, Scotland. His first scientific paper, published when he was only 14 years old, described a generalized series of oval curves that could be traced with pins and thread by analogy with an ellipse. This fascination with geometry and with mechanical models continued throughout his career and was of great help in his subsequent research. At the age of 16 he entered the University of Edinburgh, where he read voraciously on all subjects and published two more scientific papers. In 1850 he went to the University of Cambridge, where his exceptional powers began to be recognized.

In 1860 Maxwell was appointed to the professorship of natural philosophy at King's College, London. His early investigations led to a demonstration of colour photography through the use of red, green, and blue filters. He was elected to the Royal Society in 1861. His theoretical and experimental work on the viscosity of gases was also undertaken during these years and culminated in a lecture to the Royal Society in

1866. He supervised the experimental determination of electrical units for the British Association for the Advancement of Science, and this work in measurement and standardization led to the establishment of the National Physical Laboratory. He also measured the ratio of electromagnetic and electrostatic units of electricity and confirmed that it was in satisfactory agreement with the velocity of light as predicted by his theory on the electromagnetic field.

In 1865 Maxwell resigned his professorship at King's College and retired to the family estate, Glenlair in Scotland. During this period he was devoted to writing his famous *Treatise on Electricity and Magnetism* (1873). It was Maxwell's research on electromagnetism that established him among the great scientists of history. In the preface to his treatise, the best exposition of his theory, Maxwell stated that his major task was to convert the physical ideas of Faraday (1791–1867) into mathematical form. In attempting to illustrate Faraday's law of induction (that a changing magnetic field gives rise to an induced electromagnetic field), Maxwell constructed a mechanical model.

His theory suggested that electromagnetic waves could be generated in a laboratory – a possibility first demonstrated by Heinrich Hertz in 1887, eight years after Maxwell's death. The resulting radio industry with its many applications thus has its origin in Maxwell's publications.

In addition to his electromagnetic theory, Maxwell made major contributions to other areas of physics. While still in his twenties, he demonstrated his mastery of classical physics by writing a prizewinning essay on Saturn's rings, in which he concluded that the rings must consist of masses of matter not mutually coherent – a conclusion that was corroborated more than 100 years later by the first Voyager space probe to reach Saturn.

The Maxwell relations of equality between different partial derivatives of thermodynamic functions are included in every standard textbook on thermodynamics. Though Maxwell did not originate the modern kinetic theory of gases, he was the first to apply the methods of probability and statistics in describing the properties of an assembly of molecules. Thus he was able to demonstrate that the velocities of molecules in a gas, previously assumed to be equal, must follow a statistical distribution (known subsequently as the Maxwell-Boltzmann distribution law). In later papers Maxwell investigated the transport properties of gases – i.e. the effect of changes in temperature and pressure on viscosity, thermal conductivity, and diffusion.

In 1871 Maxwell was elected to the new Cavendish professorship at the University of Cambridge. He set about designing the Cavendish Laboratory and supervised its construction.

Maxwell was far from being an abstruse theoretician. He was skilful in the design of experimental apparatus, as was shown early in his career during his investigations of colour vision. He devised a colour top with adjustable sectors of tinted paper to test the three-colour hypothesis of Thomas Young (1773–1829), and later invented a colour box that made it possible to conduct experiments with spectral colours rather than pigments. His investigations of the colour theory led him to conclude that a colour photograph could be produced by photographing through filters of the three primary colours and then recombining the images.

DMITRY IVANOVICH MENDELEYEV
(1834–1907)

Russian chemist who developed the
periodic classification of the elements.

Mendeleyev was born in the small Siberian town of Tobolsk
as the last of 14 (or 13, depending on the source) surviving
children of Ivan Pavlovich Mendeleyev, a teacher at the
local gymnasium, and Mariya Dmitriyevna Kornileva. In
1856 Mendeleyev received a master's degree and began to
conduct research in organic chemistry. Financed by a government fellowship, he went to study abroad for two years
at the University of Heidelberg. Instead of working closely
with the prominent chemists of the university, including
Robert Bunsen, Emil Erlenmeyer, and August Kekulé, he set
up a laboratory in his own apartment. In September 1860
he attended the International Chemistry Congress in Karls-
ruhe, convened to discuss such crucial issues as atomic
weights, chemical symbols, and chemical formulae. There
he met and established contacts with many of Europe's
leading chemists.

In 1861 Mendeleyev returned to St Petersburg, where he
obtained a professorship at the Technological Institute in
1864. After the defence of his doctoral dissertation in 1865
he was appointed professor of chemical technology at the
University of St Petersburg (now St Petersburg State University). He became professor of general chemistry in 1867 and
continued to teach at the university until 1890.

As he began to teach inorganic chemistry, Mendeleyev could
not find a textbook that met his needs. Since he had already
published a textbook on organic chemistry in 1861 that had
been awarded the prestigious Demidov Prize, he set out to

write another one. The result was *Osnovy Khimii* ("The Principles of Chemistry"; 1868–71), which became a classic, running through many editions and many translations. When Mendeleyev began to compose the chapter on the halogen elements (chlorine and its analogues) at the end of the first volume, he compared the properties of this group of elements to those of the group of alkali metals such as sodium. Within these two groups of dissimilar elements, he discovered similarities in the progression of atomic weights, and he wondered if other groups of elements exhibited similar properties. After studying the alkaline earths, Mendeleyev established that the order of atomic weights could be used not only to arrange the elements within each group but also to arrange the groups themselves. Thus, in his effort to make sense of the extensive knowledge that already existed of the chemical and physical properties of the chemical elements and their compounds, Mendeleyev discovered the periodic law.

His newly formulated law was announced before the Russian Chemical Society in March 1869 with the statement "elements arranged according to the value of their atomic weights present a clear periodicity of properties." Mendeleyev's law allowed him to build up a systematic table of all the 70 elements then known. He had such faith in the validity of the periodic law that he proposed changes to the generally accepted values for the atomic weight of a few elements and predicted the locations within the table of unknown elements together with their properties. At first the periodic system did not raise interest among chemists. However, with the discovery of the predicted elements, notably gallium in 1875, scandium in 1879, and germanium in 1886, it began to win wide acceptance. Gradually the periodic law and periodic table became the framework for a great part of chemical theory. By the time Mendeleyev died in 1907, he enjoyed international recognition and had received distinctions and awards from many countries.

Since Mendeleyev is best known today as the discoverer of the periodic law, his chemical career is often viewed as a long process of maturation of his main discovery. However, one striking feature of his extensive career is the diversity of his activities, including a variety of contributions to the field of physical chemistry. He conducted a broad research programme throughout his career that focused on gases and liquids. In 1860, while working in Heidelberg, he defined the "absolute point of ebullition" (the point at which a gas in a container will condense to a liquid solely by the application of pressure). In 1864 he formulated a theory (subsequently discredited) that solutions are chemical combinations in fixed proportions. In 1871, as he published the final volume of the first edition of his *Principles of Chemistry*, he was investigating the elasticity of gases and gave a formula for their deviation from Boyle's law (now also known as the Boyle-Mariotte law, the principle that the volume of a gas varies inversely with its pressure). In the 1880s he studied the thermal expansion of liquids.

Mendeleyev was one of the founders of the Russian Chemical Society (now the Mendeleyev Russian Chemical Society) in 1868 and published most of his later papers in its journal. He was a prolific thinker and writer. His published works include 400 books and articles, and to this day numerous unpublished manuscripts are kept in the Dmitry Mendeleyev Museum and Archives at St Petersburg State University. In addition, in order to supplement his income he started writing articles on popular science and technology for journals and encyclopaedias as early as 1859.

His interest in spreading scientific and technological knowledge was such that he continued popular science writing until the end of his career, taking part in the project of the *Brockhaus Enzyklopädie* and launching a series of publications entitled *Biblioteka promyshlennykh znany* ("Library of Industrial Knowledge") in the 1890s. Another interest, that of developing the agricultural and industrial resources of Russia, began to occupy Mendeleyev in the 1860s and grew to become one of his major preoccupations. In March 1890 he had to resign his chair at the university as a result of his support for protesting students, and he started a second career. He first acted as a government consultant until he was appointed director of the Central Bureau of Weights and Measures, created in 1893. There he made significant contributions to metrology. Refusing to content himself solely with the managerial aspect of his position (which involved the renewal of the prototypes of length and weight and the determination of standards), he purchased expensive precision instruments, enlarged the team of the bureau, and conducted extensive research on metrology. He was thereby able to combine his lifetime interests in science and industry and to achieve one of his main goals: integrating Russia into the western world.

ROBERT KOCH (1843–1910)

German physician and
one of the founders of bacteriology.

Koch was born in Clausthal, Hannover (now Clausthal-Zellerfeld, Germany). He attended the University of Göttingen, where he studied medicine, graduating in 1866. He then became a physician in various provincial towns. After serving briefly as a field surgeon during the Franco–Prussian War of 1870–1, he became district surgeon in Wollstein, where he built a small laboratory. Equipped with a microscope, a microtome (an instrument for cutting thin slices of tissue), and a home-made incubator, he began his study of algae, switching later to pathogenic (disease-causing) organisms.

One of Koch's teachers at Göttingen had been the anatomist and histologist Friedrich Gustav Jacob Henle, who in 1840 had published the theory that infectious diseases are caused by living microscopic organisms. In 1850 the French parasitologist Casimir Joseph Davaine was among the first to observe organisms in the blood of diseased animals. In 1863 he reported the transmission of anthrax by the inoculation of healthy sheep with the blood of animals dying of the disease, and the finding of microscopic rod-shaped bodies in the blood of both groups of sheep. The natural history of the disease was, nevertheless, far from complete.

It was at that point that Koch began. He cultivated the anthrax organisms in suitable media on microscope slides, demonstrated their growth into long filaments, and discovered the formation within them of oval, translucent bodies – dormant spores. Koch found that the dried spores could remain viable for years, even under exposed conditions. The finding explained the recurrence of the disease in pastures long unused for grazing, for the dormant spores could, under the right conditions, develop into the rod-shaped bacteria (bacilli) that cause anthrax. Koch's discovery of the anthrax life cycle was announced and illustrated at Breslau in 1876, on the invitation of Ferdinand Cohn, an eminent botanist. Julius Cohnheim, a famous pathologist, was deeply impressed by Koch's presentation. "It leaves nothing more to be proved," he said.

In 1877 Koch published an important paper on the investigation, preservation, and photographing of bacteria. His work was illustrated by superb photomicrographs. In his paper he described his method of preparing thin layers of bacteria on glass slides and fixing them by gentle heat. Koch also invented the apparatus and the procedure for the very useful hanging-drop technique, whereby microorganisms could be cultured in a drop of nutrient solution on the underside of a glass slide. The following year he summarized his experiments on the etiology (causation) of wound infection. By inoculating animals with material from various sources, he produced six types of infection, each caused by a specific microorganism. He then transferred these infections by inoculation through several kinds of animals, reproducing the original six types. In that study, he observed differences in pathogenicity for different species of hosts and demonstrated that the animal body is an excellent apparatus for the cultivation of bacteria.

Koch, now recognized as a scientific investigator of the first rank, obtained a position in Berlin in the Imperial Health Office, where he set up a laboratory in bacteriology. With his collaborators, he devised new research methods to isolate pathogenic bacteria. Koch determined guidelines to prove that a disease is caused by a specific organism. These four criteria, called Koch's postulates, are: (1) A specific microorganism is always associated with a given disease, (2) The microorganism can be isolated from the diseased animal and grown in pure culture in the laboratory, (3) The cultured microbe will cause disease when transferred to a healthy animal, and (4) The same type of microorganism can be isolated from the newly infected animal.

Koch concentrated his efforts on the study of tuberculosis, with the aim of isolating its cause. Although it was suspected that tuberculosis was caused by an infectious agent, the organism had not yet been isolated and identified. By modifying the method of staining, Koch discovered the tubercle bacillus and established its presence in the tissues of animals and humans suffering from the disease. A fresh difficulty arose when for some time it proved impossible to grow the organism in pure culture. But eventually Koch succeeded in isolating the organism in a succession of media and induced tuberculosis in animals by inoculating them with it. Its etiologic role was thereby established. On 24 March 1882, Koch announced before the Physiological Society of Berlin that he had isolated and grown the tubercle bacillus, which he believed to be the cause of all forms of tuberculosis.

Meanwhile, Koch's work was interrupted by an outbreak of cholera in Egypt and the danger of its transmission to Europe. As a member of a German government commission, Koch went to Egypt to investigate the disease. Although he soon had reason to suspect a particular comma-shaped bacterium (*Vibrio cholerae*) as the cause of cholera, the epidemic ended before he was able to confirm his hypothesis. Proceeding to India, where cholera is endemic, he completed his task, identifying both the organism responsible for the disease and its transmission via drinking water, food, and clothing.

Not an eloquent speaker, Koch was nevertheless by example, demonstration, and precept one of the most effective of teachers, and his numerous pupils – from the entire western world and Asia – were the creators of the new era of bacteriology. His work on trypanosomes was of direct use to the eminent German bacteriologist Paul Ehrlich (1854–1915) – just one example of Koch's instigation of epochal work both within and beyond his own immediate sphere. His discoveries and his technical innovations were matched by his theoretical contribution of the fundamental concepts of disease etiology.

Long before his death, his place in the history of science was universally recognized.

GEORG CANTOR (1845–1918)

German mathematician who founded set theory
and introduced the mathematically meaningful
concept of transfinite numbers.

Cantor was born in St Petersburg, Russia. In 1863, after briefly attending the University of Zürich, he transferred to the University of Berlin to specialize in physics, philosophy, and mathematics. There he was taught by the mathematicians Karl Theodor Weierstrass (1815–97), whose field of expertise, analysis, probably had the greatest influence on him; Ernst Eduard Kummer, in higher arithmetic; and Leopold Kronecker, a specialist on the theory of numbers who later opposed him. Following one semester at the University of Göttingen in 1866, Cantor wrote his doctoral thesis in 1867, *In re mathematica ars propendi pluris facienda est quam solvendi* ("In Mathematics the Art of Asking Questions is More Valuable than Solving Problems"), on a question that Carl Friedrich Gauss had left unsettled in his *Disquisitiones Arithmeticae* (1801). After a brief teaching assignment in a Berlin girls' school, Cantor joined the faculty at the University of Halle, where he remained for the rest of his life.

In a series of ten papers from 1869 to 1873, Cantor dealt first with the theory of numbers. On the suggestion of Heinrich Eduard Heine, a colleague at Halle who recognized his ability, Cantor then turned to the theory of trigonometric series, in which he extended the concept of real numbers. Starting from

the work done by the German mathematician Bernhard Riemann in 1854, Cantor in 1870 showed that the function of a
complex variable could be represented in only one way by a
trigonometric series. This led him to begin his major lifework,
the theory of sets and the concept of transfinite numbers.

An important exchange of letters with Richard Dedekind,
mathematician at the Brunswick Technical Institute, who was
his lifelong friend and colleague, marked the beginning of
Cantor's ideas on the theory of sets. Both agreed that a set,
whether finite or infinite, is a collection of objects (e.g. the
integers $\{0, \pm 1, \pm 2, \ldots\}$) that share a particular property,
while each object retains its own individuality. But when
Cantor applied the device of the one-to-one correspondence
(e.g. $\{a, b, c\}$ to $\{1, 2, 3\}$) to study the characteristics of sets, he
quickly saw that they differed in the extent of their membership, even among infinite sets. (A set is infinite if one of its
parts, or subsets, has as many objects as itself.) His method
soon produced surprising results.

In 1873 Cantor demonstrated that the rational numbers,
though infinite, are countable because they may be placed in a
one-to-one correspondence with the natural numbers (i.e. the
integers, as $1, 2, 3, \ldots$). He showed that the set (or aggregate)
of real numbers (composed of irrational and rational numbers)
was infinite and uncountable. Even more paradoxically, he
proved that the set of all algebraic numbers contains as many
components as the set of all integers and that transcendental
numbers (those that are not algebraic, as pi), which are a
subset of the irrationals, are uncountable and are therefore
more numerous than integers, which must be conceived as
infinite.

But Cantor's paper, in which he first put forward these
results, was refused for publication in *Crelle's Journal* by one
of its referees, Kronecker, who henceforth vehemently opposed

Cantor's work. On Dedekind's intervention, however, it was published in 1874 as "Über eine Eigenschaft des Inbegriffes aller reellen algebraischen Zahlen" ("On a Characteristic Property of All Real Algebraic Numbers").

Cantor's theory became a whole new subject of research concerning the mathematics of the infinite (e.g. an endless series, as 1, 2, 3, . . ., and even more complicated sets), and his theory was heavily dependent on the device of the one-to-one correspondence. In thus developing new ways of asking questions concerning continuity and infinity, Cantor quickly became controversial.

In 1895–7 Cantor fully propounded his view of continuity and the infinite, including infinite ordinals and cardinals, in his best known work, *Beiträge zur Begründung der transfiniten Mengelehre* (published in English under the title "Contributions to the Founding of the Theory of Transfinite Numbers"; 1915). This work contains his conception of transfinite numbers, to which he was led by his demonstration that an infinite set may be placed in a one-to-one correspondence with one of its subsets. By the smallest transfinite cardinal number he meant the cardinal number of any set that can be placed in one-to-one correspondence with the positive integers. This transfinite number he referred to as aleph-null. Larger transfinite cardinal numbers were denoted by aleph-one, aleph-two, . . . He then developed an arithmetic of transfinite numbers that was analogous to finite arithmetic. Thus, he further enriched the concept of infinity. The opposition he faced and the length of time before his ideas were fully assimilated represented in part the difficulties of mathematicians in reassessing the ancient question: "What is a number?" Cantor demonstrated that the set of points on a line possessed a higher cardinal number than aleph-null. This led to the famous problem of the continuum hypothesis, namely, that there are no cardinal numbers between aleph-null and the cardinal number of the points on a line. This problem has, in the first and second halves of the twentieth century, been of great interest to the mathematical world and was studied by many mathematicians, including the Czech-Austrian-American Kurt Gödel (1906–78) and the American Paul J. Cohen (1934–2007).

Although mental illness, beginning in about 1884, afflicted the last years of his life, Cantor remained actively at work. In 1897 he helped to convene in Zürich the first International Congress of Mathematicians. Partly because he had been opposed by Kronecker, he often sympathized with young, aspiring mathematicians and sought to find ways to ensure that they would not suffer, as he had, because of entrenched faculty members who felt threatened by new ideas. At the turn of the century, his work was fully recognized as fundamental to the development of function theory, of analysis, and of topology. Moreover, his work stimulated further development of both the intuitionist and the formalist schools of thought in the logical foundations of mathematics; it has substantially altered mathematical education in the United States and is often associated with the "new mathematics".

HENRI POINCARÉ (1854–1912)

French mathematician, one of the
greatest mathematicians and mathematical
physicists at the end of nineteenth century.

Poincaré was born in Nancy, France. He studied mathematics from 1873 to 1875 at the École Polytechnique in Paris and continued his studies at the Mining School in Caen before receiving his doctorate from the École Polytechnique in 1879. While a student, Poincaré showed promise when he discovered new types of complex functions that solved a wide variety of differential equations. This major work involved one of the first "mainstream" applications of non-Euclidean geometry, a subject discovered by the Hungarian János Bolyai and the Russian Nikolay Lobachevsky in about 1830 but not generally accepted by mathematicians until the 1860s and '70s. Poincaré published a long series of papers on this work in 1880–84 that effectively made his name internationally.

In the 1880s Poincaré also began work on curves defined by a particular type of differential equation. He investigated such questions as: Do the solutions spiral into or away from a point? Do they, like the hyperbola, at first approach a point and then swing past and recede from it? Do some solutions form closed loops? If so, do nearby curves spiral toward or away from these closed loops? He showed that the number and types of singular points are determined purely by the topological nature of the surface. In particular, it is only on the torus (a doughnut-shaped surface) that the differential equations he was considering have no singular points.

Poincaré intended this preliminary work to lead to the study of the more complicated differential equations that describe the motion of the solar system. In 1885 an added inducement to take the next step presented itself when King Oscar II of Sweden offered a prize for anyone who could establish the stability of the solar system. This would require showing that equations of motion for the planets could be solved and the orbits of the planets shown to be curves that stay in a bounded region of space for all time. Some of the greatest mathematicians since Isaac Newton had attempted to solve this problem, and Poincaré soon realized that he could not make any headway unless he concentrated on a simpler, special case, in which two massive bodies orbit one another in circles around their common centre of gravity while a minute third body orbits them both. The third body is taken to be so small that it does not affect the orbits of the larger ones.

Poincaré could establish that the orbit is stable, in the sense that the small body returns infinitely often arbitrarily close to any position it has occupied. For this and other achievements in his essay, Poincaré was awarded the prize in 1889. But, on writing the essay for publication, he discovered that another result in it was wrong, and in putting that right he discovered even small changes in the initial conditions could produce large, unpredictable changes in the resulting orbit. (This phenomenon is now known as pathological sensitivity to initial positions, and is one of the characteristic signs of a chaotic system). Poincaré summarized his new mathematical methods in astronomy in his three-volume *Les Méthodes nouvelles de la mécanique céleste* ("The New Methods of Celestial Mechanics"; 1892, 1893, 1899).

Poincaré was led by this work to contemplate mathematical spaces (now called manifolds) in which the position of a point is determined by several coordinates. Very little was known about such manifolds, and, although the German mathematician Bernhard Riemann (1826–66) had hinted at them a generation or more earlier, few had taken the hint. Poincaré took up the task and looked for ways in which such manifolds could be distinguished, thus opening up the whole subject of topology. He singled out the idea of considering closed curves in the manifold that cannot be deformed into one another. For example, any curve on the surface of a sphere can be continuously shrunk to a point, but there are curves on a torus (curves wrapped around a hole, for instance) that cannot. Poincaré asked if a three-dimensional manifold in which every curve can be shrunk to a point is topologically equivalent to a three-dimensional sphere. This problem (now known as the Poincaré conjecture) became one of the most important problems in algebraic topology.

Poincaré's main achievement in mathematical physics was his magisterial treatment of the electromagnetic theories of Hermann von Helmholtz (1821–94), Heinrich Hertz (1857–1894), and Hendrik Lorentz (1853–1928). His interest in this topic – which, he showed, seemed to contradict Newton's laws of mechanics – led him to write a paper in 1905 on the motion of the electron. This paper, and others of his at this time, came close to anticipating the discovery by Albert Einstein, in the same year, of the theory of special relativity.

In about 1900 Poincaré acquired the habit of writing up accounts of his work in the form of essays and lectures for the general public. Published as *La Science et l'Hypothèse* ("Science and Hypothesis", 1903), *La Valeur de la science* ("The Value of Science"; 1905), and *Science et méthode* ("Science and Method"; 1908), these essays form the core of his reputation as a philosopher of mathematics and science. His most famous claim in this connection is that much of science is a matter of convention.

In many ways Poincaré's influence was extraordinary. All the topics discussed above led to the creation of new branches of mathematics that are still highly active today, and he also contributed a large number of more technical results. Yet in other ways his influence was slight. He never attracted a group of students around him, and the younger generation of French mathematicians tended to keep him at a respectful distance. His failure to appreciate Einstein helped to relegate his work in physics to obscurity after the revolutions of special and general relativity. His often imprecise mathematical exposition, masked by a delightful prose style, was alien to the generation of the 1930s who modernized French mathematics under the collective pseudonym of Nicolas Bourbaki, and they proved to be a powerful force. However, the diversity and fecundity of his work has begun to prove attractive again in a world that sets more store by applicable mathematics and less by systematic theory.

SIGMUND FREUD (1856–1939)

Austrian neurologist, the founder of psychoanalysis.

Freud was born in Freiberg, Moravia, in the Austrian Empire (now Příbor, Czech Republic). His father, Jakob, was a Jewish wool merchant who had been married once before he wed Freud's mother, Amalie Nathansohn. The father, 40 years old at Freud's birth, seems to have been a relatively remote and authoritarian figure, while his mother appears to have been more nurturing and emotionally available. Although Freud had two older half-brothers, his strongest (if also most am bivalent) attachment seems to have been to a nephew, John, one year his senior, who provided the model of intimate friend and hated rival that Freud reproduced often at later stages of his life.

In 1859 the Freud family was compelled for economic reasons to move to Leipzig and then a year later to Vienna, where Freud remained until the Nazi annexation of Austria, 78 years later. Despite Freud's dislike of the imperial city, in part because of its citizens' frequent anti-Semitism, his psycho-analysis reflected in significant ways the cultural and political context out of which it emerged.

In 1873 Freud was graduated from the Sperl Gymnasium

and, apparently inspired by a public reading of an essay by Goethe on nature, turned to medicine as a career. At the University of Vienna he worked with one of the leading physiologists of his day, Ernst von Brücke. In 1882 he entered the General Hospital in Vienna as a clinical assistant to train with the psychiatrist Theodor Meynert, and in 1885 Freud was appointed lecturer in neuropathology, having concluded important research on the brain's medulla. Freud's scientific training remained of cardinal importance in his work, or at least in his own conception of it.

In late 1885 Freud left Vienna to continue his studies of neuropathology at the Salpêtrière clinic in Paris, where he worked under the guidance of Jean-Martin Charcot. His 19 weeks in the French capital proved a turning point in his career, for Charcot's work with patients classified as "hysterics" introduced Freud to the possibility that psychological disorders might have their source in the mind rather than the brain. He returned to Vienna in February 1886 with the seed of his revolutionary psychological method implanted.

Back in Vienna, Freud began his closest friendship with the Berlin physician Wilhelm Fliess, whose role in the development of psychoanalysis has occasioned widespread debate. Throughout the 15 years of their friendship Fliess provided Freud with an invaluable interlocutor for his most daring ideas.

A somewhat less controversial influence arose from the partnership Freud began with the physician Josef Breuer. Freud had turned to a clinical practice in neuropsychology, and the office he established at Berggasse 19 was to remain his consulting room for almost half a century. During the early 1880s, Breuer had treated a patient named Bertha Pappenheim – or "Anna O.," as she became known in the literature – who was suffering from a variety of hysterical symptoms. Rather than using hypnotic suggestion, Breuer allowed her to lapse into a state resembling autohypnosis, in which she would talk about the initial manifestations of her symptoms. To Breuer's surprise, the very act of verbalization seemed to provide some relief. "The talking cure" or "chimney sweeping", as Breuer and Anna O., respectively, called it, seemed to act cathartically to produce an abreaction, or discharge, of the pent-up emotional blockage at the root of the pathological behaviour. Freud did not grasp the full implications of Breuer's experience until a decade later, when he developed the technique of free association. This revolutionary method was announced in the work Freud published jointly with Breuer in 1895, *Studien über Hysterie* ("Studies in Hysteria"). By encouraging the patient to express any random thoughts that came associatively to mind, the technique aimed at uncovering hitherto unarticulated material from the realm of the psyche that Freud, following a long tradition, called the unconscious.

Difficulty in freely associating – sudden silences, stuttering, or the like – suggested to Freud the importance of the material struggling to be expressed, as well as the power of what he called the patient's defences against that expression. Such blockages Freud dubbed resistance, which had to be broken down in order to reveal hidden conflicts. Unlike Charcot and Breuer, Freud came to the conclusion, based on his clinical experience with female "hysterics", that the most insistent source of resisted material was sexual in nature. And, even more momentously, he linked the etiology (causation) of neurotic symptoms to the same struggle between a sexual feeling or urge and the psychic defences against it. Being able to bring that conflict to consciousness through free association and then probing its implications was thus a crucial step, he reasoned, on the road to relieving the symptom – which was best understood as an unwitting compromise formation between the wish and the defence.

At first, however, Freud was uncertain about the precise status of the sexual component in this dynamic conception of the psyche. His patients seemed to recall actual experiences of early seductions, often incestuous in nature. Freud's initial impulse was to accept these as having happened. But then, as he disclosed in a now-famous letter to Fliess of September 2 1897, he concluded that, rather than being memories of actual events, these shocking recollections were the residues of infantile impulses and desires to be seduced by an adult. What was recalled was not a genuine memory but what he would later call a screen memory, or fantasy, hiding a primitive wish. Freud concluded that the fantasies and yearnings of the child were at the root of later conflict.

The absolute centrality of his change of heart in the subsequent development of psychoanalysis cannot be doubted. For in attributing sexuality to children, emphasizing the causal power of fantasies, and establishing the importance of repressed desires, Freud laid the groundwork for what many have called the epic journey into his own psyche.

Freud's work on hysteria had focused on female sexuality and its potential for neurotic expression. To be fully universal, psychoanalysis – a term Freud coined in 1896 – would also have to examine the male psyche in a condition of what might be called normality. It would have to become more than a psychotherapy and develop into a complete theory of the mind. To this end Freud accepted the enormous risk of generalizing from the experience he knew best: his own. Significantly, his self-analysis was both the first and the last in the history of the movement he spawned: all future analysts would have to undergo a training analysis with someone whose own analysis was ultimately traceable to Freud's analysis of his disciples.

Freud's self-exploration was apparently enabled by a disturbing event in his life. In October 1896, Jakob Freud died shortly before his 81st birthday. Emotions were released in his son that he understood as having been long repressed – emotions concerning his earliest familial experiences and feelings. Beginning in earnest in July 1897, Freud attempted to reveal their meaning by drawing on a technique that had been available for millennia: the deciphering of dreams. Freud's contribution to the tradition of dream analysis was pathbreaking, for in insisting on them as "the royal road to a knowledge of the unconscious," he provided a remarkably elaborate account of why dreams originate and how they function. In what many commentators consider his master work, *Die Traumdeutung* ("The Interpretation of Dreams"; published in 1899, but given the date of the dawning century to emphasize its epochal character), he presented his findings.

Dreams are the disguised expression of wish fulfilments, Freud said. Like neurotic symptoms, they are the effects of compromises in the psyche between desires and prohibitions in conflict with their realization. Although sleep can relax the power of the mind's diurnal censorship of forbidden desires, such censorship nonetheless persists in part during nocturnal existence. Dreams, therefore, have to be decoded to be understood, and not merely because they are actually forbidden desires experienced in distorted fashion. For dreams undergo further revision in the process of being recounted to the analyst.

In 1904 Freud published *Zur Psychopathologie des Alltagslebens* ("The Psychopathology of Everyday Life"), in which he explored such seemingly insignificant errors as slips of the tongue or pen (later colloquially called Freudian slips), misreadings, or forgetting of names. These errors Freud understood to have symptomatic and thus interpretable importance. But, unlike dreams, they need not betray a repressed infantile wish, yet can arise from more immediate hostile, jealous, or egoistic causes.

In 1905 Freud extended the scope of this analysis by examining *Der Witz und seine Beziehung zum Unbewussten* ("Jokes and Their Relation to the Unconscious"). Invoking the idea of "joke-work" as a process comparable to dreamwork, he also acknowledged the double-sided quality of jokes, at once consciously contrived and unconsciously revealing. Seemingly innocent phenomena such as puns or jests are as open to interpretation as more obviously tendentious, obscene, or hostile jokes. The explosive response often produced by successful humour, Freud contended, owes its power to the orgasmic release of unconscious impulses, aggressive as well as sexual. But insofar as jokes are more deliberate than dreams or slips, they draw on the rational dimension of the psyche, which Freud was to call the "ego", as much as on what he was to call the "id". In addition to the neurosis of hysteria, with its conversion of affective conflicts into bodily symptoms, Freud developed complicated etiological explanations for other typical neurotic behaviour, such as obsessive-compulsions, paranoia, and narcissism. These he called psychoneuroses, because of their rootedness in childhood conflicts, as opposed to the actual neuroses such as hypochondria, neurasthenia, and anxiety neurosis, which are due to problems in the present (the last, for example, being caused by the physical suppression of sexual release).

In *Jenseits des Lustprinzips* ("Beyond the Pleasure Principle"; 1920) and *Das Ich und das Es* ("The Ego and the Id"; 1923), Freud attempted to clarify the relationship between his earlier topographical division of the psyche into the unconscious, preconscious, and conscious, and his subsequent structural categorization into id, ego, and superego. The id was defined in terms of the most primitive urges for gratification in the infant – urges dominated by the desire for pleasure and the cathexis (the concentration in one direction) of energy. Governed by no laws of logic, indifferent to the demands of expediency, and unconstrained by the resistance of external reality, the id is ruled by what Freud called "the primary process directly expressing somatically generated instincts". Through the inevitable experience of frustration the infant learns to adapt itself to the exigencies of reality. The secondary process that results leads to the growth of the ego, which follows what Freud called "the reality principle", in contra-distinction to "the pleasure principle" dominating the id. Here the need to delay gratification in the service of self-preservation is slowly learned in an effort to thwart the anxiety produced by unfulfilled desires. What Freud termed "defence mechanisms" are developed by the ego to deal with such conflicts. Repression is the most fundamental defence mechanism, but Freud also posited an entire repertoire of others, including reaction formation, isolation, undoing, denial, displacement, and rationalization. The last component in Freud's trichotomy, the superego, develops from the internalization of society's moral commands through identification with parental dictates during the resolution of the Oedipus complex. Only partly conscious, the superego gains some of its punishing force by borrowing certain aggressive elements in the id, which are turned inward against the ego and produce feelings of guilt. But it is largely through the internalization of social norms that the superego is constituted – an acknowledgement that prevents psychoanalysis from conceptualizing the psyche in purely biologistic or individualistic terms.

Freud's final major work, *Der Mann Moses und die mono-theistische Religion* ("Moses and Monotheism"; 1938), was more than just the "historical novel" he had initially thought to subtitle it. Moses had long been a figure of capital importance for Freud; indeed Michelangelo's famous statue of Moses had been the subject of an essay written in 1914. The book itself sought to solve the mystery of Moses's origins by claiming that he was actually an aristocratic Egyptian by birth who had chosen the Jewish people to keep alive an earlier monotheistic religion. Too stern and demanding a taskmaster, Moses was slain in a Jewish revolt, and a second, more pliant leader, also called Moses, rose in his place. The guilt engendered by the parricidal act was, however, too much to endure, and the Jews ultimately returned to the religion given them by the original Moses as the two figures were merged into one in their memories. Here Freud's ambivalence about his religious roots and his father's authority was allowed to pervade a highly fanciful story that reveals more about its author than its ostensible subject.

Moses and Monotheism was published in the year Hitler invaded Austria. Freud was forced to flee to England. Freud himself died only a few weeks after World War II broke out, at a time when his worst fears about the irrationality lurking behind the facade of civilization were being realized. Freud's death did not, however, hinder the reception and dissemination of his ideas, and a plethora of Freudian schools emerged to develop psychoanalysis in different directions. In fact, despite the relentless and often compelling challenges mounted against virtually all of his ideas, Freud has remained one of the most potent figures in the intellectual landscape.

NIKOLA TESLA (1856–1943)

Serbian-American inventor and engineer who discovered and patented the rotating magnetic field.

Tesla was born in Smiljan, Austria-Hungary (now in Croatia). Training for an engineering career, he attended the Technical

University at Graz, Austria, and the University of Prague. At Graz he first saw the Gramme dynamo, which operated as a generator and, when reversed, became an electric motor, and he conceived a way to use alternating current to advantage. Later, at Budapest, he visualized the principle of the rotating magnetic field and developed plans for an induction motor that would become his first step toward the successful utilization of alternating current.

In 1882 Tesla went to work in Paris for the Continental Edison Company, and, while on assignment to Strassburg (now Strasbourg), Alsace, in 1883, he constructed, in afterwork hours, his first induction motor. Tesla sailed for America in 1884 – arriving in New York with four cents in his pocket, a few of his own poems, and calculations for a flying machine. He first found employment with Thomas Edison (1847–1931), but the two inventors were far apart in background and methods, and their separation was inevitable.

In May 1885, George Westinghouse, head of the Westinghouse Electric Company in Pittsburgh, bought the patent rights to Tesla's polyphase system of alternating-current dynamos, transformers, and motors. The transaction precipitated a titanic power struggle between Edison's direct-current systems and the Tesla–Westinghouse alternating-current approach, which eventually won out.

Tesla soon established his own laboratory, where his inventive mind could be given free rein. In order to allay fears about alternating currents, Tesla gave exhibitions in his laboratory in which he lit lamps by allowing electricity to flow through his body. He was often invited to lecture at home and abroad. The Tesla coil, which he invented in 1891, became widely used today in radio and television sets and other electronic equipment.

Westinghouse used Tesla's alternating current system to light the World's Columbian Exposition at Chicago in 1893. This success was a factor in their winning the contract to install the first power machinery at Niagara Falls, which bore Tesla's name and patent numbers. The project carried power to Buffalo by 1896.

In Colorado Springs, Colorado, where he stayed from May 1899 until early 1900, Tesla made what he regarded as his most important discovery – terrestrial stationary waves. By this discovery he proved that the Earth could be used as a conductor and made to resonate at a certain electrical frequency. He also lit 200 lamps without wires from a distance of 40 km (25 miles) and created man-made lightning, producing flashes measuring 41 metres (135 feet). At one time he was certain he had received signals from another planet in his Colorado laboratory, a claim that was met with derision in some scientific journals.

Returning to New York in 1900, Tesla began construction on Long Island of a wireless world broadcasting tower, with $150,000 capital from the American financier J. Pierpont Morgan. Tesla claimed he secured the loan by assigning 51 per cent of his patent rights of telephony and telegraphy to Morgan. He expected to provide worldwide communication and to furnish facilities for sending pictures, messages, weather warnings, and stock reports. However, the project was abandoned because of a financial panic, labour troubles, and Morgan's withdrawal of support. It was Tesla's greatest defeat.

Tesla's work then shifted to turbines and other projects. Because of a lack of funds, his ideas remained in his notebooks, which are still examined by enthusiasts for unexploited clues. In 1917 Tesla was the recipient of the Edison Medal, the highest honour that the American Institute of Electrical Engineers could bestow.

MAX PLANCK (1858–1947)

German theoretical physicist who
originated quantum theory.

Planck was born in Kiel, Schleswig, Germany, the sixth child
of a distinguished jurist and professor of law at the University
of Kiel. He entered the University of Munich in the autumn of
1874 but found little encouragement. During a year spent at
the University of Berlin in 1877–8, he was similarly unim-
pressed. His intellectual capacities were, however, brought to
a focus as the result of his independent study, especially of
Rudolf Clausius' writings on thermodynamics. Returning to
Munich, he received his doctoral degree in July 1879 at the
unusually young age of 21. In 1885, with the help of his
father's professional connections, he was appointed *ausser-
ordentlicher Professor* ("associate professor") at the University of Kiel. In 1889 he received an appointment to the
University of Berlin and in 1892 was promoted to *ordentlicher Professor* ("full professor"). His Berlin lectures on all
branches of theoretical physics went through many editions
and exerted great influence. He remained in Berlin for the rest
of his active life.

Planck recalled that his "original decision to devote myself
to science was a direct result of the discovery . . . that the laws
of human reasoning coincide with the laws governing the
sequences of the impressions we receive from the world about
us; that, therefore, pure reasoning can enable man to gain an
insight into the mechanism of the [world] . . ." He deliberately
decided, in other words, to become a theoretical physicist at a
time when theoretical physics was not yet recognized as a
discipline in its own right.

The first instance of an absolute in nature that impressed

Planck deeply, even as a *Gymnasium* student, was the law of the conservation of energy – the first law of thermodynamics. Later, during his university years, he became equally convinced that the entropy law, the second law of thermodynamics, was also an absolute law of nature. The second law became the subject of his doctoral dissertation at Munich, and it lay at the core of the researches that led him to discover the quantum of action, now known as Planck's constant, h, in 1900.

By the 1890s various experimental and theoretical attempts had been made to determine the spectral energy distribution (the curve showing how much radiant energy is emitted at different frequencies for a given temperature) of a "black-body" – an object that re-emits all of the radiant energy incident upon it. Planck was particularly attracted to the formula found in 1896 by his colleague Wilhelm Wien, and he subsequently made a series of attempts to derive "Wien's law" on the basis of the second law of thermodynamics. By October 1900, however, other colleagues had found definite indications that Wien's law, while valid at high frequencies, broke down completely at low frequencies.

Planck knew how the entropy of the radiation had to depend mathematically upon its energy in the high-frequency region if Wien's law held there. He also saw what this dependence had to be in the low-frequency region in order to reproduce the experimental results there. Planck guessed, therefore, that he should try to combine these two expressions in the simplest way possible, and to transform the result into a formula relating the energy of the radiation to its frequency.

The result, which is known as Planck's radiation law, led him to the realization that in the world of atomic dimensions energy could not be absorbed or emitted continuously but only in discrete amounts, or quanta, of energy. Planck's concept of energy quanta conflicted fundamentally with all past physical theory and showed that the microphysical world could not in principle be described by classical mechanics. Planck was driven to introduce the concept strictly by the force of his logic: he was, as one historian put it, a reluctant revolutionary.

Indeed, it was years before the far-reaching consequences of Planck's achievement were generally recognized, and in this Albert Einstein (1879–1955) played a central role. In 1905, independently of Planck's work, Einstein argued that under certain circumstances radiant energy itself seemed to consist of quanta (light quanta, later called photons), and in 1907 he showed the generality of the quantum hypothesis by using it to interpret the temperature dependence of the specific heats of solids. Henri Poincaré (1854–1912) later provided a mathematical proof that Planck's radiation law necessarily required the introduction of quanta. In 1913 Niels Bohr also contributed greatly to its establishment through his quantum theory of the hydrogen atom.

Ironically, Planck himself was one of the last to struggle for a return to classical theory – a stance he later regarded not with regret but as a means by which he had thoroughly convinced himself of the necessity of the quantum theory. Opposition to Einstein's radical light quantum hypothesis of 1905 persisted until after the discovery of the Compton effect (the decrease in energy of photons of X-rays or other electromagnetic radiation when they interact with matter) in 1922.

Planck was 42 years old in 1900 when he made the famous discovery, which won him the 1918 Nobel Prize for Physics and brought him many other honours. It is not surprising that he subsequently made no discoveries of comparable importance. He was, however, the first prominent physicist to champion Einstein's special theory of relativity in 1905.

Planck became permanent secretary of the mathematics and physics sections of the Prussian Academy of Sciences in 1912, and held that position until 1938; he was also president of the Kaiser Wilhelm Society (now the Max Planck Society) from 1930 to 1937. These offices and others placed him in a position of great authority, especially among German physicists.

WILLIAM BATESON (1861–1926)

English biologist who founded
and named the science of genetics.

William Bateson was born in Whitby, Yorkshire. He obtained his master's degree from the University of Cambridge and was throughout his career a dedicated evolutionist. In 1894 he published *Materials for the Study of Variation*, which stated that evolution could not occur through a continuous variation of species, since distinct features often appeared or disappeared suddenly in plants and animals. Realizing that discontinuous variation could be understood only after something was known about the inheritance of traits, Bateson began work on the experimental breeding of plants and animals.

In 1900, he discovered an article, "Experiments with Plant Hybrids", written by Gregor Mendel, an Austrian monk, 34 years earlier. The paper, found in the same year by Hugo de Vries, Carl Correns, and Erich Tschermak von Seysenegg, dealt with the appearance of certain features in successive generations of garden peas. Bateson noted that his breeding results were explained perfectly by Mendel's paper and that the monk had succinctly described the transmission of elements governing heritable traits in his plants.

Bateson translated Mendel's paper into English, and during the next ten years became Mendel's champion in England, corroborating his principles experimentally. He published, with Reginald Punnett, the results of a series of breeding experiments in 1905–8 that not only extended Mendel's principles to animals (poultry) but also showed that certain features were consistently inherited together, apparently counter to Mendel's findings. This phenomenon, which came to be termed linkage, is now known to be the result of the occurrence of genes located in close proximity on the same chromosome. Bateson's experiments also demonstrated a dependence of certain characters on two or more genes. Unfortunately, he misinterpreted his results, refusing to accept the interpretation of linkage advanced by the geneticist Thomas Hunt Morgan (1866–1945). In fact, he opposed Morgan's entire chromosome theory, advocating his own "vibratory" theory of inheritance, founded on laws of force and motion – a concept that found little acceptance among other scientists.

Bateson became, at the University of Cambridge, the first British professor of genetics (1908). He left this chair in 1910 to spend the rest of his life directing the John Innes Horticultural Institution at Merton, South London (later moved to Norwich), transforming it into a centre for genetic research. His books include *Mendel's Principles of Heredity* (1902) and *Problems of Genetics* (1913).

MARIE CURIE (1867–1934) AND PIERRE CURIE (1859–1906)

Polish-born French physicist and her husband, a French physicist, both noted for their work on radioactivity.

Maria Sklodowska was born in Warsaw, Poland (then part of the Russian Empire). From childhood she was remarkable for her prodigious memory, and at the age of 16 she won a gold medal on completion of her secondary education at the Russian lycée. In 1891 she went to Paris and, now using the name Marie, began to follow the lectures of Paul Appel, Gabriel Lippmann, and Edmond Bouty at the Sorbonne. Sklodowska worked far into the night in her student-quarters garret and virtually lived on bread and butter and tea. She came first in the *licence* of physical sciences in 1893. She began to work in Lippmann's research laboratory, and in 1894 came second in the *licence* of mathematical sciences. It was in the spring of that year that she met Pierre Curie. Pierre, who was born in Paris, was carrying out research on magnetism for his doctoral thesis in physics and had previously conducted important studies on crystals.

Their marriage, on 25 July 1895, marked the start of a partnership that was soon to achieve results of world significance, in particular the discovery of polonium (so called by Marie in honour of her native land) in the summer of 1898, and of radium a few months later. Following the discovery by Henri Becquerel in 1896 of a new phenomenon (which she later called radioactivity), Marie Curie, looking for a subject for a thesis, decided to find out if the property discovered in uranium was to be found in other matter. She discovered that this was true for thorium at the same time as did the German physicist G.C. Schmidt.

Turning her attention to minerals, Curie found her interest drawn to pitchblende, a mineral whose activity, superior to that of pure uranium, could be explained only by the presence in the ore of small quantities of an unknown substance of very high activity. Pierre Curie then joined her in the work that she undertook to resolve this problem and which led to the discovery of the new elements polonium and radium. While Pierre devoted himself chiefly to the physical study of the new elements, Marie struggled to obtain pure radium in the metallic state. On the results of this research, she received her doctorate of science in June 1903 and, with Pierre, was awarded the Davy Medal of the Royal Society. Also in 1903 they shared with Becquerel the Nobel Prize for Physics for the discovery of radioactivity.

The birth of her two daughters, Irène and Ève, in 1897 and 1904 did not interrupt Curie's intensive scientific work. She was appointed lecturer in physics at the École Normale Supérieure for girls in Sèvres in 1900, and introduced there a method of teaching based on experimental demonstrations. In December 1904 she was appointed chief assistant in the laboratory directed by Pierre Curie.

The sudden death of Pierre on 19 April 1906 was a bitter blow to Marie Curie, but it was also a decisive turning point in her career: henceforth she was to devote all her energy to completing alone the scientific work that they had undertaken. On 13 May 1906 she was appointed to the professorship that had been left vacant on her husband's death; she was the first woman to teach in the Sorbonne. In 1908 she became titular professor, and in 1910 her fundamental treatise on radioactivity was published. In 1911 she was awarded the Nobel Prize for Chemistry, for the isolation of pure radium. In 1914 she saw the completion of the building of the laboratories of the Institut du Radium (Radium Institute) at the University of Paris.

Throughout World War I, Curie, with the help of her daughter Irène, devoted herself to the development of the use of X-radiography. In 1918 the Radium Institute, the staff of which Irène had joined, began to operate in earnest, and it was to become a universal centre for nuclear physics and chemistry. Curie, now at the highest point of her fame and, from 1922, a member of the Academy of Medicine, devoted her research to the study of the chemistry of radioactive substances and the medical applications of these substances.

One of Curie's outstanding achievements was to have understood the need to accumulate intense radioactive sources, not only to treat illness but also to maintain an abundant supply for research in nuclear physics; the resultant stockpile was an unrivalled instrument until the appearance after 1930 of particle accelerators. The existence in Paris at the Radium Institute of a stock of 1.5 grams (0.05 ounces) of radium made a decisive contribution to the success of the experiments undertaken in the years around 1930, and in particular of those performed by Irène Curie in conjunction with Frédéric Joliot. This work prepared the way for the discovery of the neutron by Sir James Chadwick (1891–1974) and, above all, for the discovery in 1934 by Irène and Frédéric Joliot-Curie of artificial radioactivity.

A few months after this discovery, Marie Curie died as a result of leukaemia caused by the action of radiation. Her contribution to physics was immense, not only in her own work, but also in her influence on subsequent generations of nuclear physicists and chemists.

SIR ERNEST RUTHERFORD, 1ST BARON RUTHERFORD OF NELSON (1871–1937)

British physicist who laid the groundwork
for the development of nuclear physics.

Rutherford was the fourth of the twelve children of James, a wheelwright at Brightwater near Nelson on South Island, New Zealand, and Martha Rutherford. His parents, who had emigrated from Great Britain, denied themselves many comforts so that their children might be well educated.

On his arrival in Cambridge in 1895, Rutherford began to work under Sir J.J. Thomson, Professor of Experimental Physics at the university's Cavendish Laboratory. Continuing his work on the detection of Hertzian waves over a distance of 3.2 km (2 miles), he gave an experimental lecture on his results before the Cambridge Physical Society and was delighted when his paper was published in the *Philosophical Transactions* of the Royal Society, a signal honour for so young an investigator.

Rutherford made a great impression on colleagues in the Cavendish Laboratory, and Thomson held him in high esteem. He also aroused jealousies in the more conservative members of the Cavendish fraternity. In December 1895, when Wilhelm Röntgen discovered X-rays, Thomson asked Rutherford to join him in a study of the effects of passing a beam of X-rays through a gas. They discovered that the X-rays produced large quantities of electrically charged particles, and that these ionized atoms recombined to form neutral molecules. Working on his own, Rutherford then devised a technique for measuring the velocity and rate of recombination of these positive and negative ions. The papers that the two published on this subject remain classics to this day.

In 1896 the French physicist Henri Becquerel discovered that uranium emitted rays that could fog a photographic plate as did X-rays. Rutherford soon showed that they also ionized air but that they were different from X-rays, consisting of two distinct types of radiation. He named them alpha rays, highly powerful in producing ionization but easily absorbed; and beta rays, which produced less radiation but had more penetrating ability.

Within three years Rutherford succeeded in making important advances in an entirely new area of physics called radioactivity. He soon discovered that thorium or its compounds disintegrated into a gas that in turn disintegrated into an unknown "active deposit", also radioactive. Rutherford and a young chemist, Frederick Soddy, then investigated three groups of radioactive elements – radium, thorium, and actinium. They concluded in 1902 that radioactivity was a process in which atoms of one element spontaneously disintegrated into atoms of an entirely different element, which also remained radioactive. This interpretation was opposed by many chemists, who held firmly to the concept of the indestructibility of matter; the suggestion that some atoms could tear themselves apart to form entirely different kinds of matter was to them a remnant of medieval alchemy. Nevertheless, Rutherford's outstanding work won him recognition by the Royal Society, which elected him a fellow in 1903 and awarded him the Rumford medal in 1904. Rutherfod summarized the results of his research in his book *Radio-activity* (1904).

A prodigious worker with tremendous powers of concentration, Rutherford continued to make a succession of brilliant discoveries – and with remarkably simple apparatus. For example, he showed in 1903 that alpha rays can be deflected by electric and magnetic fields (the direction of the deflection proving that the rays are particles of positive charge), and he determined their velocity and the ratio of their charge to their mass. These results were obtained by passing such particles between thin, matchbox-sized metal plates stacked closely together: in one experiment each plate was charged oppositely to its neighbour; in another the assembly was placed in a strong magnetic field. In each experiment he measured the strengths of the fields, which just sufficed to prevent the particles from emerging from the stack.

With his student Thomas D. Royds, Rutherford proved in 1908 that the alpha particle really is a helium atom, by allowing alpha particles to escape through the thin glass wall of a containing vessel into an evacuated outer glass tube, and showing that the spectrum of the collected gas was that of helium. Almost immediately, in 1908, he was awarded a Nobel Prize – but for chemistry, for his investigations concerning the disintegration of elements.

In 1911 Rutherford made his greatest contribution to science, with his nuclear theory of the atom. He had observed that fast-moving alpha particles on passing through thin plates of mica produced diffuse images on photographic plates, whereas a sharp image was produced when there was no obstruction to the passage of the rays. He considered that the particles must be deflected through small angles as they passed close to atoms of the mica, but calculation showed that an electric field of 100,000,000 volts per centimetre was necessary to deflect such particles travelling at 20,000 km per second (12,400 miles per second) – a most astonishing conclusion.

Rutherford suggested to Hans Geiger, a research assistant, and Ernest Marsden, a student, that it would be of interest to examine whether any particles were scattered backward – i.e. deflected through an angle of more than 90 degrees. To their astonishment, a few particles in every 10,000 were indeed so scattered, emerging from the same side of a gold foil as that on which they had entered. After a number of calculations, Rutherford came to the conclusion that the intense electric field required to cause such a large deflection could occur only if all the positive charge in the atom, and therefore almost all the mass, were concentrated on a very small central nucleus some 10,000 times smaller in diameter than that of the entire atom. The positive charge on the nucleus would therefore be balanced by an equal charge on all the electrons distributed somehow around the nucleus. This theory of atomic structure is known as the Rutherford atomic model. A knighthood, conferred in 1914, further marked the public recognition of Rutherford's services to science.

During World War I Rutherford worked on the practical problem of submarine detection by underwater acoustics. He produced the first artificial disintegration of an element in 1919, when he found that on collision with an alpha particle an atom of nitrogen was converted into an atom of oxygen and an atom of hydrogen. That same year he succeeded Thomson as Cavendish professor.

Rutherford's service as president of the Royal Society (1925–30) and as chairman of the Academic Assistance Council, which helped almost 1,000 university refugees from Germany, increased the claims upon his time. But whenever possible he worked in the Cavendish Laboratory, where he encouraged students, probed for the facts, and always sought an explanation in simple terms.

CARL GUSTAV JUNG (1875–1961)

Swiss psychologist and psychiatrist
who founded analytic psychology.

Jung was born in Kesswil, Switzerland, and was the son of a philologist and pastor. He seemed destined to become a minister, for there were a number of clergymen on both sides of his family. In his teens he discovered philosophy and read widely, and chose to study medicine and become a psychiatrist. He was a student at the universities of Basel (1895–1900) and then Zürich, where he gained his master's degree in 1902.

Jung was fortunate in joining the staff of the Burghölzli Asylum of the University of Zürich in 1900, when it was under the direction of Eugen Bleuler, whose psychological interests had initiated what are now considered classical studies of mental illness. At Burghölzli, Jung began, with outstanding success, to apply association tests initiated by earlier researchers. He studied in particular patients' peculiar and illogical responses to stimulus words, and found that they were caused by emotionally charged clusters of associations withheld from consciousness because of their disagreeable, immoral (to them), and frequently sexual content. He used the term "complex" to describe such conditions.

These researches, which established Jung as a psychiatrist of international repute, led him to understand the investigations of Freud (1856–1939); his findings confirmed many of Freud's ideas, and, for a period of five years between 1907 and 1912, he was Freud's close collaborator. He held important positions in the psychoanalytic movement and was widely thought of as the most likely successor to the founder of psychoanalysis. But this was not to be the outcome of their relationship. Partly for temperamental reasons and partly because of differences of viewpoint, the collaboration ended. At this stage Jung differed with Freud largely over the latter's insistence on the sexual bases of neurosis. A serious disagreement came in 1912, with the publication of Jung's *Wandlungen und Symbole der Libido* ("Psychology of the Unconscious"; 1916), which ran counter to many of Freud's ideas. Although Jung had been elected president of the International Psychoanalytic Society in 1911, he resigned from the society in 1914.

Jung's first achievement was to differentiate two classes of people according to attitude types: extraverted (outward-looking) and introverted (inward-looking). Later he differentiated four functions of the mind – thinking, feeling, sensation, and intuition – one or more of which predominate in any given person. Results of this study were embodied in *Psychologische Typen* ("Psychological Types"; 1921). Jung's wide scholarship was well manifested here, as it had been in *Psychology of the Unconscious*.

As a boy Jung had remarkably striking dreams and powerful fantasies that had developed with unusual intensity. After his break with Freud, he deliberately allowed this aspect of himself to function again and gave the irrational side of his nature free expression. At the same time, he studied it scientifically by keeping detailed notes of his strange experiences. He later developed the theory that these experiences came from an area of the mind that he called the collective unconscious, which he held was shared by everyone. This much-contested conception was combined with a theory of archetypes that Jung held as fundamental to the study of the psychology of religion. In Jung's terms, archetypes are instinctive patterns, have a universal character, and are expressed in behaviour and images.

Jung devoted the rest of his life to developing his ideas,

especially those on the relationship between psychology and religion. In his view, obscure and often neglected texts of writers in the past shed unexpected light not only on Jung's own dreams and fantasies but also on those of his patients. He thought it necessary for the successful practice of psychotherapists' art that they become familiar with writings of the old masters.

Besides the development of new psychotherapeutic methods that derived from his own experience, and the theories developed from them, Jung gave fresh importance to the so-called Hermetic tradition. He conceived that the Christian religion was part of a historic process necessary for the development of consciousness, and he also thought that the heretical movements, starting with Gnosticism and ending in alchemy, were manifestations of unconscious archetypal elements not adequately expressed in the mainstream forms of Christianity. He was particularly impressed with his finding that alchemical-like symbols could frequently be found in modern dreams and fantasies, and he thought that alchemists had constructed a kind of textbook of the collective unconscious. He expounded on this in four out of the eighteen volumes that make up his *Collected Works*.

Jung's historical studies aided him in pioneering the psychotherapy of the middle-aged and elderly, especially those who felt their lives had lost meaning. He helped them to appreciate the place of their lives in the sequence of history. Most of these patients had lost their religious belief; Jung found that if they could discover their own myth as expressed in dream and imagination they would become more complete personalities. He called this process "individuation".

In later years he became professor of psychology at the Federal Polytechnical University in Zürich (1933–41) and professor of medical psychology at the University of Basel (1943). His personal experience, his continued psychotherapeutic practice, and his wide knowledge of history placed him in a unique position to comment on current events. As early as 1918 he had begun to think that Germany held a special position in Europe; the Nazi revolution was, therefore, highly significant for him, and he delivered a number of hotly contested views that led to his being wrongly branded as a Nazi sympathizer.

ALBERT EINSTEIN (1879–1955)

German-born physicist who developed the special and general theories of relativity.

Einstein was born in Ulm, Württemberg, Germany, to secular, middle-class Jews. His father, Hermann Einstein, was originally a featherbed salesman and later ran an electrochemical factory with moderate success. His mother, the former Pauline Koch, ran the family household.

Einstein wrote that two "wonders" deeply affected his early years. The first was his encounter with a compass at the age of five: he was mystified that invisible forces could deflect the needle, and this would lead to a lifelong fascination with such forces. The second came at the age of 12 when he discovered a book of geometry, which he devoured, calling it his "sacred little geometry book."

Einstein became deeply religious at this time, even composing several songs in praise of God and chanting religious songs on the way to school. This began to change, however, after he read science books that contradicted his religious beliefs. This challenge to established authority left a deep and lasting impression. At the Luitpold Gymnasium, Einstein often felt out of place and victimized by a Prussian-style educational system that seemed to stifle originality and creativity.

Another important influence on Einstein was a young medical student, Max Talmud (later Max Talmey), who often had dinner at the Einstein home. Talmud became an informal tutor, introducing Einstein to higher mathematics and philosophy. A pivotal point occurred when Talmud introduced him to a children's science series by Aaron Bernstein, Naturwissenschaftliche Volksbucher ("Popular Books on Physical Science"; 1867–8), in which the author imagined riding alongside electricity that was travelling inside a telegraph wire. Einstein then asked himself the question that would dominate his thinking for the next ten years: What would a light beam look like if you could run alongside it? If light were a wave, then the light beam should appear stationary, like a frozen wave. Even as a child, however, he knew that stationary light waves had never been observed, so there was a paradox.

Einstein's education was disrupted by his father's repeated failures at business. In 1894 Hermann Einstein moved to Milan, Italy, to work with a relative, and the young Albert was left at a boarding house in Munich and expected to finish his education. Alone, miserable, and repelled by the looming prospect of military duty when he turned 16, Einstein ran away six months later and arrived on the doorstep of his surprised parents. As a school dropout and draft dodger with no employable skills, his prospects did not look promising.

Fortunately, Einstein could apply directly to the Eidgenössische Polytechnische Schule (Swiss Federal Polytechnic School) in Zürich without the equivalent of a high school diploma if he passed its stiff entrance examinations. His marks showed that he excelled in mathematics and physics, but he failed at French, chemistry, and biology. Because of his exceptional maths scores, he was allowed into the polytechnic on the condition that he first finish his formal schooling. He went to a special high school run by Jost Winteler in Aarau, Switzerland, and graduated in 1896. He also renounced his German citizenship at that time – he was stateless until 1901, when he was granted Swiss citizenship.

Einstein became lifelong friends with the Winteler family, with whom he had been boarding. He met many students who would become loyal friends, such as Marcel Grossmann, a mathematician; and Michele Besso, with whom he enjoyed lengthy conversations about space and time. He also met his future wife, Mileva Maric, a fellow physics student from Serbia. His parents vehemently opposed the relationship, but Einstein defied them, and he and Mileva even had a child, Lieserl, in January 1902, whose fate is unknown (it is commonly thought that she died of scarlet fever or was given up for adoption).

In 1902 Einstein reached perhaps the lowest point in his life. He could not marry Mileva and support a family without a job, and his father's business went bankrupt. Einstein took lowly jobs tutoring children, but he was fired from even these. Later that year, however, Grossman's father recommended him for a position as a clerk in the Swiss patent office in Bern. Then Einstein's father became seriously ill and, just before he died, gave his blessing for his son to marry Mileva. They married in January 1903 and had two children, Hans Albert and Eduard.

In retrospect, Einstein's job at the patent office was a blessing. He would quickly finish analysing patent applications, leaving him time to daydream about the vision that had obsessed him since he was 16: what would happen if you race alongside a light beam? While at the polytechnic school he had studied the equations of James Clerk Maxwell (1831–79), which describe the nature of light, and discovered a fact unknown to Maxwell himself – namely, that the speed of light remained the same no matter how fast one moved. This violated the laws of motion of Isaac Newton (1642–1727), however, because there is no absolute velocity in Newton's theory. This insight led Einstein to formulate the principle of relativity: "the speed of light is a constant in any inertial [uniformly moving] frame."

During 1905, often called Einstein's "miracle year", he published four papers in the *Annalen der Physik* ("Annals of Physics"), each of which would alter the course of modern physics. In the first, "Über einen die Erzeugung und Verwandlung des Lichtes betreffenden heuristischen Gesichtspunkt" ("On a Heuristic Viewpoint Concerning the Production and Transformation of Light"), Einstein applied the quantum theory to light in order to explain the photoelectric effect. If light occurs in tiny packets (later called photons), he said, then it should knock out electrons in a metal in a precise way. In "Über die von der molekularkinetischen Theorie der Wärme geforderte Bewegung von in ruhenden Flüssigkeiten suspendierten Teilchen" ("On the Movement of Small Particles Suspended in Stationary Liquids Required by the Molecular-Kinetic Theory of Heat"), Einstein offered the first experimental proof of the existence of atoms. By analysing the motion of tiny particles suspended in still water, called Brownian motion, he could calculate the size of the jostling atoms and Avogadro's number (the number of atoms in 12 grams of carbon-12; this number of atoms or molecules is termed a mole). In "Zur Elektrodynamik bewegter Körper" ("On the Electrodynamics of Moving Bodies"), Einstein laid out the mathematical theory of special relativity. And in "Ist die Trägheit eines Körpers von seinem Energieinhalt abhängig?" ("Does the Inertia of a Body Depend Upon Its Energy Content?"), submitted almost as an afterthought, he showed that special relativity theory led to the equation $E = mc^2$. This states that there is an equivalence between energy and mass: that energy equals mass times the speed of light squared. It provided the first mechanism to explain the energy source of the sun and other stars.

Other scientists, especially Henri Poincaré (1854–1912) and Hendrik Lorentz (1853–1928), had formulated pieces of the theory of special relativity, but Einstein was the first to assemble the whole theory and to realize that it was a universal law of nature, not a curious figment of motion in the ether, as Poincaré and Lorentz had thought. (In one private letter to his wife, Einstein referred to "our theory", which has led some to speculate that she was a cofounder of relativity theory. However, Mileva had abandoned physics after twice failing her graduate exams, and there is no record of her involvement in developing relativity. In his 1905 paper, Einstein credits only his conversations with Besso in developing relativity.)

In the nineteenth century there were two pillars of physics: Newton's laws of motion and Maxwell's theory of light. Einstein was alone in realizing that they were in contradiction and that one of them must fall.

At first Einstein's 1905 papers were ignored by the physics community. This began to change after he received the attention of just one physicist – perhaps the most influential physicist of his generation – Max Planck, the founder of quantum theory. Soon, owing to Planck's laudatory comments and to experiments that gradually confirmed his theories, Einstein was invited to lecture at international meetings, such as the Solvay Conferences, and he rose rapidly in the academic world. He was offered a series of positions at increasingly prestigious institutions, including the University of Zürich, the

University of Prague, the Swiss Federal Institute of Technology, and finally the University of Berlin, where he served as director of the Kaiser Wilhelm Institute for Physics from 1913 to 1933 (although the opening of the institute was delayed until 1917).

As his fame spread, Einstein's marriage was falling apart. He was constantly on the road, speaking at international conferences, and lost in contemplation of relativity. Convinced that his marriage was doomed, Einstein began an affair with a cousin, Elsa Löwenthal, whom he later married. When he finally divorced Mileva in 1919, he agreed to give her the money he might receive if he ever won a Nobel Prize.

One of the deep thoughts that consumed Einstein from 1905 to 1915 was a crucial flaw in his own theory: it made no mention of gravitation or acceleration. For the next ten years he would be absorbed with formulating a theory of gravity in terms of the curvature of space-time. To Einstein, Newton's gravitational force was actually a by-product of a deeper reality – the bending of the fabric of space and time. This was the essence of his general theory of relativity, which he completed in November 1915 and considered to be his masterpiece.

In the summer of that year, Einstein gave six two-hour lectures at the University of Göttingen that thoroughly explained general relativity, albeit with a few unfinished mathematical details. Much to his consternation, however, the mathematician David Hilbert, who had organized the lectures at his university, then completed these details and submitted a paper in November on general relativity just five days before Einstein, as if the theory were his own. Later the two resolved their differences and remained friends. Today physicists refer to the equations as the Einstein-Hilbert action, but the theory itself is attributed solely to Einstein.

Einstein was convinced that general relativity was correct

because of its mathematical beauty and because it accurately predicted the perihelion (point of closest approach of orbit around the sun) of Mercury. His theory also predicted a measurable deflection of light around the sun. As a consequence, he even offered to help fund an expedition to measure the deflection of starlight during an eclipse of the sun.

Einstein's work was interrupted by World War I. A lifelong pacifist, he was one of only four intellectuals in Germany to sign a manifesto opposing the country's entry into war. Disgusted, he called nationalism "the measles of mankind". He would write, "At such a time as this, one realizes what a sorry species of animal one belongs to."

After the war, two expeditions were sent to test Einstein's prediction of deflected starlight near the sun. One set sail for the island of Principe, off the coast of West Africa, and the other to Sobral in northern Brazil to observe the solar eclipse of 29 May 1919. On 6 November 1919, the results, which confirmed Einstein's prediction, were announced in London at a joint meeting of the Royal Society and the Royal Astronomical Society. The headline of *The Times* of London read, "Revolution in Science – New Theory of the Universe – Newtonian Ideas Overthrown.'" Almost immediately, Einstein became a world-renowned physicist, the successor to Isaac Newton.

Invitations came pouring in for him to speak around the world. In 1921 he began the first of several world tours, visiting the United States, England, Japan, and France. Everywhere he went, the crowds numbered in the thousands. En route from Japan, he received word that he had received the Nobel Prize for Physics, but for the photoelectric effect rather than for his relativity theories. During his acceptance speech, Einstein startled the audience by speaking about relativity instead of the photoelectric effect.

Einstein also launched the new science of cosmology. His equations predicted that the universe is dynamic – expanding or contracting. This contradicted the prevailing view that the universe was static, so he reluctantly introduced a "cosmological term" to stabilize his model of the universe. In 1929 the astronomer Edwin Hubble found that the universe was indeed expanding, thereby confirming Einstein's earlier work. In 1930, in a visit to the Mount Wilson Observatory near Los Angeles, Einstein met with Hubble and declared the cosmological constant to be his "greatest blunder". Recent satellite data, however, have shown that the cosmological constant is probably not zero but actually dominates the matter-energy content of the entire universe. Einstein's "blunder" apparently determines the ultimate fate of the universe.

Einstein also clarified his religious views, stating that he believed there was an "old one" who was the ultimate lawgiver. He wrote that he did not believe in a personal God that intervened in human affairs, but rather in the God of the seventeenth-century Dutch Jewish philosopher Benedict de Spinoza (1632–77) – the God of harmony and beauty. Einstein's task, he believed, was to formulate a master theory that would allow him to "read the mind of God." He would write:

> I'm not an atheist and I don't think I can call myself a pantheist. We are in the position of a little child entering a huge library filled with books in many different languages . . . The child dimly suspects a mysterious order in the arrangement of the books but doesn't know what it is. That, it seems to me, is the attitude of even the most intelligent human being toward God.

*   *   *

In December 1932 Einstein decided to leave Germany forever. It became obvious to him that his life was in danger: a Nazi organization had published a magazine with Einstein's picture and the caption "Not Yet Hanged" on the cover. There was even a price on his head. So great was the threat that Einstein split with his pacifist friends and said that it was justified to defend yourself with arms against Nazi aggression. He settled in the United States at the newly formed Institute for Advanced Study at Princeton, New Jersey, which soon became a mecca for physicists from around the world. Newspaper articles declared that the "pope of physics" had left Germany and that Princeton had become the new Vatican.

To Einstein's horror, during the late 1930s physicists began seriously to consider whether his equation $E = mc^2$ might make an atomic bomb possible. In 1920 Einstein himself had considered but eventually dismissed the possibility. However, he left it open if a method could be found to magnify the power of the atom. Then in 1938–9 Otto Hahn, Fritz Strassmann, Lise Meitner, and Otto Frisch showed that vast amounts of energy could be unleashed by the splitting of the uranium atom. The news electrified the physics community.

Many scientists saw the perils to world peace if Hitler's scientists should apply the principle of the nuclear chain reaction to the production of an atomic bomb. After the physicists Leo Szilard and Eugene Wigner met with Einstein in July 1939 and explained the threat to him, Einstein agreed to write a letter to US president, Franklin D. Roosevelt, to warn him of the possibility of Nazi Germany's developing an atomic bomb and the danger it posed. Following several translated drafts, Einstein signed a letter on 2 August that was delivered to Roosevelt on October 11. Roosevelt wrote back on 19 October, informing Einstein that he had organized the Uranium Committee to study the issue.

Einstein was granted permanent residency in the United States in 1935 and became an American citizen in 1940, although he chose to retain his Swiss citizenship. During the war, many of his colleagues were asked to journey to the desert town of Los Alamos, New Mexico, to develop the first atomic bomb for the Manhattan Project (a US government programme to harness nuclear energy for military purposes). But Einstein, the man whose equation had set the whole effort into motion, was never asked to participate. Several thousand declassified FBI files reveal the reason: the U.S. government feared Einstein's lifelong association with peace and socialist organizations.

Einstein was on holiday when he heard the news that an atomic bomb had been dropped on Japan. Almost immediately he was part of an international effort to try to bring the atomic bomb under control, forming the Emergency Committee of Atomic Scientists.

The physics community split on the question of whether to build a hydrogen bomb. Einstein opposed its development, instead calling for international controls on the spread of nuclear technology. He also was increasingly drawn to anti-war activities and to advancing the civil rights of African Americans. In 1952 David Ben-Gurion, Israeli's premier, offered Einstein the post of President of Israel. Einstein, a prominent figure in the Zionist movement, respectfully declined.

Although Einstein continued to pioneer many key developments in the theory of general relativity – such as wormholes, higher dimensions, the possibility of time travel, the existence of black holes, and the creation of the universe – he was increasingly isolated from the rest of the physics community. Because of the huge strides made by quantum theory in unravelling the secrets of atoms and molecules, the majority

of physicists were working on the quantum theory, not relativity. Through a series of sophisticated "thought experiments", Einstein tried to find logical inconsistencies in the
quantum theory, particularly its lack of a deterministic mechanism. He would often say that "God does not play dice with
the universe."

In 1935 Einstein's most celebrated attack on the quantum
theory led to the EPR (Einstein-Podolsky-Rosen) thought
experiment. According to quantum theory, under certain
circumstances two electrons separated by huge distances
would have their properties linked, as if by an umbilical
cord. Under these circumstances, if the properties of the first
electron were measured, the state of the second electron
would be known instantly – faster than the speed of light.
This conclusion, Einstein claimed, clearly violated relativity.
(Experiments conducted since then have confirmed that the
quantum theory, rather than Einstein, was correct about the
EPR experiment. In essence, what Einstein had actually
shown was that quantum mechanics is nonlocal, i.e. random
information can travel faster than light. This does not violate
relativity, because the information is random and therefore
useless.)

The other reason for Einstein's increasing detachment from
his colleagues was his obsession, beginning in 1925, with
discovering a unified field theory – an all-embracing theory
that would unify the forces of the universe, and thereby the
laws of physics, into one framework. In his later years he
stopped opposing the quantum theory and tried to incorporate
it, along with light and gravity, into a larger unified field
theory. Gradually Einstein became set in his ways. He rarely
travelled far and confined himself to long walks around
Princeton with close associates, whom he engaged in deep
conversations about politics, religion, physics, and his unified

field theory. At the time of his death, of an aortic aneurysm, however, the theory remained unfinished.

In some sense, Einstein, instead of being a relic in his later years, may have been too far ahead of his time. The "strong force", a major piece of any unified field theory, was still a total mystery in Einstein's lifetime. Only in the 1970s and '80s did physicists begin to unravel the secret of the strong force with the quark model. Nevertheless, Einstein laid the foundations for work that continues to win Nobel Prizes for succeeding physicists. In 1993 a Nobel Prize was awarded to the discoverers of gravitation waves, predicted by Einstein. In 1995 a Nobel Prize was awarded to the discoverers of Bose-Einstein condensates (a new form of matter that can occur at extremely low temperatures). New generations of space satellites have continued to verify the cosmology of Einstein, and many leading physicists are trying to finish Einstein's ultimate dream of a "theory of everything".

ALFRED LOTHAR WEGENER
(1880–1930)

German meteorologist and geophysicist
who formulated the first complete statement
of the continental drift hypothesis.

Wegener was born in Berlin, Germany. Although he earned a PhD degree in astronomy from the University of Berlin in 1905, he had an interest in paleoclimatology, and in 1906–08 he took part in an expedition to Greenland to study polar air circulation. He made three more expeditions to Greenland, in 1912–13, 1929, and 1930. He taught meteorology at Marburg and Hamburg and was a professor of meteorology and geophysics at the University of Graz from 1924 to 1930.

Like certain other scientists before him, Wegener became impressed with the similarity in the coastlines of eastern South America and western Africa, and speculated that those lands had once been joined together. In about 1910 he began toying with the idea that in the Late Paleozoic era (about 250 million years ago) all the present-day continents had formed a single large mass, or supercontinent, which had subsequently broken apart. Wegener called this ancient continent Pangaea. Other scientists had proposed such a continent but had explained the separation of the modern world's continents as having resulted from the subsidence, or sinking, of large portions of the supercontinent to form the Atlantic and Indian oceans. Wegener, by contrast, proposed that Pangaea's constituent portions had slowly moved thousands of kilometres apart over long periods of geologic time. His term for this movement was *die Verschiebung der Kontinente* ("continental displacement"), which gave rise to the term continental drift.

Wegener first presented his theory in lectures in 1912 and published it in full in 1915 in his most important work, *Die Entstehung der Kontinente und Ozeane* ("The Origin of Continents and Oceans"). He searched the scientific literature for geological and paleontological evidence that would buttress his theory, and he was able to point to many closely related fossil organisms and similar rock strata that occurred on widely separated continents, particularly those found in both the Americas and in Africa. Wegener's theory of continental drift won some adherents in the ensuing decade, but his postulations of the driving forces behind the continents' movement seemed implausible. By 1930 his theory had been rejected by most geologists, and it sank into obscurity for the next few decades, only to be resurrected as part of the theory of plate tectonics during the 1960s. Wegener died during his last expedition to Greenland in 1930.

SIR ALEXANDER FLEMING (1881–1955)

Scottish bacteriologist whose discovery of penicillin prepared the way for the highly effective practice of antibiotic therapy for infectious diseases.

Fleming was born in Lochfield, Ayr, Scotland. After taking his degree at St Mary's Hospital Medical School, London University, in 1906, he conducted experiments to discover antibacterial substances that would be nontoxic to human tissues. He continued his research while serving with distinction in the Royal Army Medical Corps in World War I. In 1918 he returned to research and teaching at St Mary's, where he was professor of bacteriology from 1928 to 1948, when he became professor emeritus.

In 1921 Fleming identified and isolated lysozyme, an enzyme found in certain animal tissues and secretions, such as tears and saliva, that exhibits antibiotic activity. While working with *Staphylococcus* bacteria in 1928, Fleming noticed a bacteria-free circle around a mould growth (spores of *Penicillium notatum*) that was contaminating a culture of the staphylococci. Investigating, he found a substance in the mould that prevented growth of the bacteria even when it was diluted 800 times.

He called it penicillin. Fleming found that penicillin is nontoxic but that it inhibits the growth of many types of disease-causing bacteria. He was aware of the significance of his discovery, but he lacked the necessary chemical means to isolate and identify the active compound involved. However, he obtained enough penicillin to use on humans topically for skin and eye infections. It was not until 11 years later in 1939, during World War II, that the pressing need for new antibacterial drugs provided the impetus for Ernst Boris Chain and Walter Florey to extend Fleming's basic discovery to the isolation, purification, testing, and production in quantity of penicillin. In 1945 Fleming – together with Chain and Florey – received the Nobel Prize for Physiology or Medicine. Fleming was elected a fellow of the Royal Society in 1943 and knighted in 1944.

NIELS BOHR (1885–1962)

Danish physicist who was the first to apply
the quantum theory to the problem of
atomic and molecular structure.

Bohr was born in Copenhagen, Denmark. He distinguished himself at the University of Copenhagen, winning a gold medal from the Royal Danish Academy of Sciences and Letters, and in 1911 he received his doctorate for a thesis on the electron theory of metals that stressed the inadequacies of classical physics for treating the behaviour of matter at the atomic level. He then went to England, intending to continue this work with Sir J. J. Thomson at the University of Cambridge. Thomson never showed much interest in Bohr's ideas on electrons in metals, however, although he had worked on this subject in earlier years. Bohr moved to Manchester in March 1912 and joined Ernest Rutherford and his group studying the structure of the atom.

At Manchester Bohr worked on the theoretical implications of the nuclear model of the atom recently proposed by Rutherford and known as the Rutherford atomic model. Bohr was among the first to see the importance of the atomic number, which indicates the position of an element in the periodic table and is equal to the number of natural units of electric charge on the nuclei of its atoms. He recognized that the various physical and chemical properties of the elements depend on the electrons moving around the nuclei of their atoms, and that only the atomic weight and possible radio-active behaviour are determined by the small but massive nucleus itself.

Rutherford's nuclear atom was both mechanically and electromagnetically unstable, but Bohr imposed stability on it by introducing the new and not-yet-clarified ideas of the quantum theory being developed by Max Planck, Albert Einstein, and other physicists. Departing radically from classical physics, Bohr postulated that any atom could exist only in a discrete set of stable or stationary states, each characterized by a definite value of its energy. This description of atomic structure is known as the Bohr atomic model.

The most impressive result of the Bohr atomic model was the way it accounted for the series of lines observed in the spectrum of light emitted by atomic hydrogen. He was able to determine the frequencies of these spectral lines to considerable accuracy from his theory, expressing them in terms of the charge and mass of the electron and Planck's constant. To do this, Bohr also postulated that an atom would not emit radiation while it was in one of its stable states but rather only when it made a transition between states. The frequency of the radiation so emitted would be equal to the difference in energy between those states divided by Planck's constant. This meant that the atom could neither absorb nor emit radiation continuously but only in finite steps or quantum jumps. It also meant that the various frequencies of the radiation emitted by an atom were not equal to the frequencies with which the electrons moved within the atom – a bold idea that some of Bohr's contemporaries found particularly difficult to accept. The consequences of Bohr's theory, however, were confirmed by new spectroscopic measurements and other experiments.

Bohr returned to Copenhagen from Manchester during the summer of 1912 and continued to develop his new approach to the physics of the atom. The work was completed in 1913 in Copenhagen but was first published in England. In 1916, after serving as a lecturer in Copenhagen and then in Manchester, Bohr was appointed to a professorship in his native city. The university created for Bohr a new Institute of Theoretical Physics, which opened its doors in 1921; he served as director for the rest of his life.

Through the early 1920s, Bohr concentrated his efforts on two interrelated sets of problems. He tried to develop a consistent quantum theory that would replace classical mechanics and electrodynamics at the atomic level and be adequate for treating all aspects of the atomic world. He also tried to explain the structure and properties of the atoms of all the chemical elements, particularly the regularities expressed in the periodic table and the complex patterns observed in the spectra emitted by atoms. In this period of uncertain foundations, tentative theories, and doubtful models, Bohr's work was often guided by his "correspondence principle". According to this principle, every transition process between stationary states as given by the quantum postulate can be "coordinated" with a corresponding harmonic component (of a single frequency) in the motion of the electrons as described by classical mechanics.

Bohr received the Nobel Prize for Physics in 1922, and Bohr's institute in Copenhagen became an international centre for work on atomic physics and the quantum theory. Bohr himself began to travel more widely, lecturing in many European countries and in Canada and the United States. At this time, more than any of his contemporaries, Bohr was convinced that even more radical changes in physics were still to come. During the next few years, a genuine quantum mechanics was created – the new synthesis that Bohr had been expecting. The new quantum mechanics required more than just a mathematical structure of calculating; it needed a physical interpretation. That physical interpretation came out of the intense discussions between Bohr and the steady stream of visitors to his world capital of atomic physics – discussions on how the new mathematical description of nature was to be linked with the procedures and the results of experimental physics.

Bohr expressed the characteristic feature of quantum physics in his "principle of complementarity", which "implies the impossibility of any sharp separation between the behaviour of atomic objects and the interaction with the measuring instruments which serve to define the conditions under which the phenomena appear." As a result, "evidence obtained under different experimental conditions cannot be comprehended within a single picture, but must be regarded as complementary in the sense that only the totality of the phenomena exhausts the possible information about the objects." This interpretation of the meaning of quantum physics, which implied an altered view of the meaning of physical explanation, gradually came to be accepted by the majority of physicists. The most famous and most outspoken dissenter, however, was Einstein.

During the 1930s Bohr continued to work on the epistemological problems raised by the quantum theory and also contributed to the new field of nuclear physics. He used a "liquid drop" model of the atomic nucleus, so called because it likened the nucleus to a liquid droplet, as a key step in the understanding of many nuclear processes. In particular, it played an essential part in 1939 in the understanding of nuclear fission (the splitting of a heavy nucleus into two parts, almost equal in mass, with the release of a tremendous amount of energy). Similarly, his compound-nucleus model of the atom proved successful in explaining other types of nuclear reactions.

Bohr's institute continued to be a focal point for theoretical physicists until the outbreak of World War II. The annual conferences on nuclear physics, as well as formal and informal visits of varied duration, brought virtually everyone concerned with quantum physics to Copenhagen at one time or another. Many of Bohr's collaborators in those years have written affectionately about the extraordinary spirit of the institute, where young scientists from many countries worked and played together in a light-hearted mood that concealed their absolutely serious concern with physics and with the darkening world outside.

When Denmark was occupied by the Germans in 1940, Bohr did what he could to maintain the work of his institute and to preserve the integrity of Danish culture against Nazi influences. In 1943, under threat of immediate arrest because of his Jewish ancestry and the anti-Nazi views he made no effort to conceal, Bohr, together with his wife and some other family members, was transported to Sweden by fishing boat in the dead of night by the Danish resistance movement. A few days later the British government sent an unarmed Mosquito bomber to Sweden, and Bohr was flown to England in a dramatic flight that almost cost him his life.

During the next two years, Bohr and one of his sons, Aage (who later followed in his father's footsteps as a theoretical physicist, director of the institute, and Nobel Prize winner in physics), took part in the projects for making a nuclear fission bomb. They worked in England for several months and then moved to Los Alamos, New Mexico, with a British research team. Bohr's concern about the terrifying prospects for humanity posed by such atomic weapons was evident as early as 1944, when he tried to persuade British prime minister Winston Churchill and U.S. president Franklin D. Roosevelt of the need for international cooperation in dealing with these problems. Although this appeal did not succeed, Bohr continued to argue for rational, peaceful policies, advocating an "open world" in a public letter to the United Nations in 1950.

In his last years Bohr tried to point out ways in which the idea of complementarity could throw light on many aspects of human life and thought. He had a major influence on several generations of physicists, deepening their approach to their science and to their lives. Profoundly international in spirit, Bohr was also firmly rooted in his own Danish culture. This was symbolized by his many public roles, particularly as president of the Royal Danish Academy from 1939 until his death.

ERWIN SCHRÖDINGER (1887–1961)

Austrian theoretical physicist who contributed to the wave theory of matter and to other fundamentals of quantum mechanics.

Schrödinger was born in Vienna. He entered the University of Vienna in 1906 and obtained his doctorate in 1910, upon which he accepted a research post at the university's Second

Physics Institute. He saw military service in World War I and then went to the University of Zürich in 1921, where he remained for the next six years. There, in a six-month period in 1926, at the age of 39 – a remarkably late age for original work by a theoretical physicist – he produced the papers that gave the foundations of quantum wave mechanics. In those papers he described his partial differential equation that is the basic equation of quantum mechanics.

Adopting a proposal made by Louis de Broglie in 1924 that particles of matter have a dual nature and in some situations act like waves, Schrödinger introduced a theory describing the behaviour of such a system by a wave equation that is now known as the Schrödinger equation. The solutions to Schrödinger's equation, unlike the solutions to Newton's equations, are wave functions that can only be related to the probable occurrence of physical events. The definite and readily visualized sequence of events of the planetary orbits of Newton is, in quantum mechanics, replaced by the more abstract notion of probability. (This aspect of the quantum theory made Schrödinger and several other physicists profoundly unhappy, and he devoted much of his later life to formulating philosophical objections to the generally accepted interpretation of the theory that he had done so much to create. For example, to highlight a paradox in this interpretation, he presented an imaginary experiment based on quantum theory in which a cat in a closed box would have to be alive and dead at the same time until the box was opened to observe its condition.)

In 1927 Schrödinger accepted an invitation to succeed Max Planck, the inventor of the quantum hypothesis, at the University of Berlin, and he joined an extremely distinguished faculty that included Albert Einstein. He remained at the university until 1933, at which time he reached the decision that he could no longer live in a country in which the persecution of

Jews had become a national policy. He then began a seven-year odyssey that took him to Austria, Great Britain, Belgium, the Pontifical Academy of Science in Rome, and finally, in 1940, to the Dublin Institute for Advanced Studies.

Schrödinger remained in Ireland for the next 15 years, conducting research both in physics and in the philosophy and history of science. During this period he wrote *What Is Life?* (1944), an attempt to show how quantum physics can be used to explain the stability of genetic structure. Although much of what Schrödinger had to say in this book has been modified and amplified by later developments in molecular biology, his book remains one of the most useful and profound introductions to the subject. In 1956 Schrödinger retired and returned to Vienna as professor emeritus at the university.

Of all of the physicists of his generation, Schrödinger stands out because of his extraordinary intellectual versatility. He was at home in the philosophy and literature of all of the western languages, and his popular scientific writing in English, which he had learned as a child, is among the best of its kind. His study of Ancient Greek science and philosophy, summarized in his *Nature and the Greeks* (1954), gave him both an admiration for the Greek invention of the scientific view of the world and a scepticism toward the relevance of science as a unique tool with which to unravel the ultimate mysteries of human existence. Schrödinger's own metaphysical outlook, as expressed in his last book, *Meine Weltansicht* ("My View of the World"; 1961), closely paralleled the mysticism of the Vedanta, a Hindu spiritual tradition.

Because of his exceptional gifts, Schrödinger was able in the course of his life to make significant contributions to nearly all branches of science and philosophy – an almost unique accomplishment at a time when the trend was toward increasing technical specialization in these disciplines.

SRINIVASA RAMANUJAN (1887–1920)

Indian mathematician who made pioneering
discoveries in the theory of numbers.

Ramanujan was born in Erode, Tamil Nadu, India. When he
was 15 years old he obtained a copy of George Shoobridge
Carr's *Synopsis of Elementary Results in Pure and Applied
Mathematics* (1880–6). This collection of some 6,000 theorems (none of the material was newer than 1860) aroused his
genius. Having verified the results in Carr's book, Ramanujan
went beyond it, developing his own theorems and ideas. In
1903 he secured a scholarship to the University of Madras, but
lost it the following year because he neglected all other studies
in pursuit of mathematics.

Ramanujan continued his work, without employment and
living in the poorest circumstances. After marrying in 1909 he
began a search for permanent employment that culminated in
an interview with a government official, Ramachandra Rao.
Impressed by Ramanujan's mathematical prowess, Rao supported his research for a time, but Ramanujan, unwilling to
exist on charity, obtained a clerical post with the Madras Port
Trust.

In 1911 Ramanujan published the first of his papers in the
Journal of the Indian Mathematical Society. His genius slowly
gained recognition, and in 1913 he began a correspondence
with the British mathematician Godfrey H. Hardy that led to a
special scholarship from the University of Madras and a grant
from Trinity College, University of Cambridge. Overcoming
his religious objections, Ramanujan travelled to England in
1914, where Hardy tutored him and collaborated with him in
research.

Ramanujan's knowledge of mathematics (most of which

he had worked out for himself) was startling. Although almost completely ignorant of what had been developed, his mastery of continued fractions was unequalled by any living mathematician. He worked out the Riemann series, the elliptic integrals, hypergeometric series, the functional equations of the zeta function, and his own theory of divergent series. On the other hand, the gaps in his knowledge were equally surprising. He knew nothing of doubly periodic functions, the classical theory of quadratic forms, or Cauchy's theorem, and had only the most nebulous idea of what constitutes a mathematical proof. Though brilliant, many of his theorems concerning prime numbers were completely wrong.

In England Ramanujan made further advances, especially in the partition of numbers. His papers were published in English and European journals, and in 1918 he became the first Indian to be elected to the Royal Society. In 1917 he contracted tuberculosis, but his condition improved sufficiently for him to return to India in 1919. He died the following year, generally unknown to the world at large but recognized by mathematicians as a phenomenal genius.

EDWIN POWELL HUBBLE (1889–1953)

American astronomer who is considered the founder of extragalactic astronomy and who provided the first evidence of the expansion of the universe.

Hubble was born in Marshfield, Missouri. His interest in astronomy flowered at the University of Chicago, where he was inspired by the astronomer George E. Hale. At Chicago,

Hubble earned both an undergraduate degree in mathematics and astronomy, in 1910, and a reputation as a fine boxer. Upon graduation, however, he turned away from both astronomy and athletics, preferring to study law as a Rhodes Scholar at the University of Oxford, where he obtained his bachelor's degree in 1912. He joined the Kentucky bar in 1913 but dissolved his practice soon after, finding himself bored with law. A man of many talents, he finally chose to focus these on astronomy, returning to the University of Chicago and its Yerkes Observatory in Wisconsin. After earning a PhD in astronomy in 1917 and serving in World War I, Hubble settled down to work at the Mount Wilson Observatory near Pasadena, California, and began to make discoveries concerning extragalactic phenomena.

While at Mount Wilson, during 1922–4, Hubble discovered that not all nebulae in the sky are part of the Milky Way galaxy. He found that certain nebulae contain stars called Cepheid variables, for which a correlation was already known to exist between periodicity and absolute magnitude. Using the further relationship between distance, apparent magnitude, and absolute magnitude, Hubble determined that these Cepheids are several hundred thousand light years away and thus outside the Milky Way system, and that the nebulae in which they are located are actually galaxies distinct from the Milky Way. This discovery, announced in 1924, forced astronomers to revise their ideas about the cosmos.

Soon after discovering the existence of these external galaxies, in 1926 Hubble undertook the task of classifying them according to their shapes and exploring their stellar contents and brightness patterns. In the course of this study, Hubble made his second remarkable discovery, in 1927 – that these galaxies are apparently receding from the Milky Way and that the further away they are, the faster they are receding. The implications of this discovery were immense. The universe, long considered static, was expanding; and, even more remarkably, as Hubble discovered in 1929, it was expanding in such a way that the ratio of the speed of the galaxies to their distance is a constant (now called Hubble's constant).

Although Hubble was correct that the universe was expanding, his calculation of the value of the constant was incorrect, implying that the Milky Way system was larger than all other galaxies and that the entire universe was younger than the surmised age of the Earth. Subsequent astronomers, however, revised Hubble's result and rescued his theory, creating a picture of a cosmos that has been expanding at a constant rate for 10 billion to 20 billion years.

For his achievements in astronomy Hubble received many honours and awards. Among his publications were *Red Shifts in the Spectra of Nebulae* (1934) and *The Hubble Atlas of Galaxies* (published posthumously, in 1961). Hubble remained an active observer of galaxies until his death. A sophisticated optical observatory that was launched into Earth's orbit by NASA in 1990 was named in his honour.

ENRICO FERMI (1901–1954)

Italian-born American physicist who was one of the chief architects of the nuclear age.

Fermi was born in Rome; the youngest of the three children of Alberto Fermi, a railroad employee, and Ida de Gattis. An energetic and imaginative student prodigy in high school, the young Enrico decided to become a physicist. At the age of 17 he entered the Reale Scuola Normale

Superior, which is associated with the University of Pisa. There he earned his doctorate at the age of 21 with a thesis on research with X-rays.

After a short visit to Rome, Fermi left for Germany with a fellowship from the Italian Ministry of Public Instruction to study at the University of Göttingen under the physicist Max Born, whose contributions to quantum mechanics were part of the knowledge prerequisite to Fermi's later work. He then returned to teach mathematics at the University of Florence.

In 1926 Fermi's paper on the behaviour of a perfect, hypothetical gas impressed the physics department of the University of Rome, which invited him to become a full professor of theoretical physics. Within a short time, Fermi brought together a new group of physicists, all of them in their early 20s. In 1926 he developed a statistical method for predicting the characteristics of electrons according to Pauli's exclusion principle, which suggests that there cannot be more than one subatomic particle that can be described in the same way. The Royal Academy of Italy recognized his work in 1929 by electing him to membership as the youngest member in its distinguished ranks.

This theoretical work at the University of Rome was of primary importance, but new discoveries soon prompted Fermi to turn his attention to experimental physics. In 1932 the existence of an electrically neutral particle, called the neutron, was discovered by Sir James Chadwick at the University of Cambridge. In 1934 Frédéric and Irène Joliot-Curie in France were the first to produce artificial radioactivity by bombarding elements with alpha particles, which are emitted as positively charged helium nuclei from polonium. Impressed by this work, Fermi conceived the idea of inducing artificial radioactivity by another method. He used neutrons obtained from radioactive beryllium to bombard elements and found that, when he reduced the speed of the neutrons by passing them through paraffin, the neutrons were especially effective in producing artificial radioactivity. He successfully used this method on a series of elements. When he used uranium of atomic weight 92 as the target of slow-neutron bombardment, however, he obtained puzzling radioactive substances that could not be identified.

Fermi's colleagues were inclined to believe that he had actually made a new, "transuranic" element of atomic number 93 – that, during bombardment, the nucleus of uranium had captured a neutron, thus increasing its atomic weight. Fermi did not make this claim, for he was not certain what had occurred; indeed, he was unaware that he was on the edge of a world-shaking discovery. As he modestly observed years later, "We did not have enough imagination to think that a different process of disintegration might occur in uranium than in any other element. Moreover, we did not know enough chemistry to separate the products from one another."

In 1938 Fermi was named a Nobel laureate in physics "for his identification of new radioactive elements produced by neutron bombardment and for his discovery of nuclear reaction effected by slow neutrons." He was given permission by the Fascist government of Mussolini to travel to Sweden to receive the award. As they had already secretly planned, Fermi and his wife and family left Italy, never to return, for they had no respect for Fascism.

Meanwhile, in 1938, three German scientists had repeated some of Fermi's early experiments. After bombarding uranium with slow neutrons, Otto Hahn, Lise Meitner, and Fritz Strassmann made a careful chemical analysis of the products formed. On 6 January 1939 they reported that the uranium atom had been split into several parts. Meitner, a mathematical physicist, slipped secretly out of Germany to Stockholm, where, together with her nephew Otto Frisch, she explained this new phenomenon as a splitting of the nucleus of the uranium atom into barium, krypton, and smaller amounts of other disintegration products. They sent a letter to the science journal *Nature*, which printed their report on 16 January 1939.

Meitner realized that this nuclear fission was accompanied by the release of stupendous amounts of energy by the conversion of some of the mass of uranium into energy in accordance with Einstein's mass–energy equation $E = mc^2$, where energy is equal to mass times the speed of light squared.

Fermi, apprised of this development soon after arriving in New York, saw its implications and rushed to greet Niels Bohr on his arrival in New York City. The Hahn–Meitner–Strassmann experiment was repeated at Columbia University, where, with further reflection, Bohr suggested the possibility of a nuclear chain reaction. It was agreed that the uranium-235 isotope, differing in atomic weight from other forms of uranium, would be the most effective atom for such a chain reaction.

Concerned that Hitler's scientists might apply the principle of the nuclear chain reaction to the production of an atomic bomb, the U.S. government in 1942 organized the Manhattan Project (a U.S. government programme to harness nuclear energy for military purposes) for the production of its own atomic bomb. Fermi was assigned the task of producing a controlled, self-sustaining nuclear chain reaction. He designed the necessary apparatus, which he called an atomic pile, and on 2 December 1942 led the team of scientists who, in a laboratory established in the squash court in the basement of Stagg Field at the University of Chicago, achieved the first self-sustaining chain reaction. The Manhattan Project succeeded in testing the first nuclear explosive device, at Alamogordo Air

Base in New Mexico on 16 July 1945, which was followed by the dropping of atomic bombs on Hiroshima and Nagasaki a few weeks later.

The Fermis had become American citizens in 1944. In 1946 Fermi became Distinguished-Service Professor for Nuclear Studies at the University of Chicago and also received the Congressional Medal of Merit. At the Metallurgical Laboratory of the University of Chicago, Fermi continued his studies of the basic properties of nuclear particles. He was also a consultant in the construction of the synchrocyclotron, a large particle accelerator at the University of Chicago. In 1950 he was elected a foreign member of the Royal Society.

Fermi made highly original contributions to theoretical physics, particularly to the mathematics of subatomic particles. Element number 100 was named for him, and the Enrico Fermi Award was established in his honour. He was the first recipient of this award of $25,000 in 1954.

JOHN VON NEUMANN (1903–1957)

Hungarian-born American mathematician who helped pioneer game theory and was one of the conceptual inventors of the stored-program digital computer.

Von Neumann was born in Budapest. He showed signs of genius in early childhood: he could joke in Classical Greek and, for a family stunt, could quickly memorize a page from a telephone book and recite its numbers and addresses. He commenced his intellectual career at a time when the influence of David Hilbert and his programme of establishing axiomatic foundations for mathematics was at a peak. Von Neumann's

"An Axiomatization of Set Theory" (1925) commanded the attention of Hilbert himself. From 1926 to 1927 von Neumann did postdoctoral work under Hilbert at the University of Göttingen. The work with Hilbert culminated in von Neumann's book *The Mathematical Foundations of Quantum Mechanics* (1932).

This mathematical synthesis reconciled the seemingly contradictory quantum mechanical formulations of Erwin Schrödinger and Werner Heisenberg. Von Neumann also claimed to prove that deterministic "hidden variables" cannot underlie quantum phenomena. This influential result pleased Niels Bohr and Heisenberg and played a strong role in convincing physicists to accept the indeterminacy of quantum theory. In contrast, the result dismayed Albert Einstein, who refused to abandon his belief in determinism.

By his mid-twenties, von Neumann found himself pointed out as a wunderkind at conferences and he produced a staggering succession of pivotal papers in logic, set theory, group theory, ergodic theory, and operator theory. Of all the principal branches of mathematics, it was only in topology and number theory that von Neumann failed to make an important contribution.

In 1928 von Neumann published *Zur Theorie der Gesellschaftsspiel* ("Theory of Parlour Games"), a key paper in the field of game theory. The nominal inspiration was the game of poker. Game theory focuses on the element of bluffing, a feature distinct from the pure logic of chess or the probability theory of roulette. Although von Neumann knew of the earlier work of the French mathematician Émile Borel (1871–1956), he gave the subject mathematical substance by proving the mini-max theorem. This asserts that for every finite, two-person zero-sum game, there is a rational outcome in the sense that two perfectly logical adversaries can arrive at a mutual choice of game strategies, confident that they could not expect to do better by choosing another strategy. In games such as poker, the optimal strategy incorporates a chance element. Poker players must bluff occasionally – and unpredictably – in order to avoid exploitation by a savvier player.

In 1929 von Neumann was asked to lecture on quantum theory at Princeton University. This led to an appointment as visiting professor from 1930 to 1933. In 1933 he became one of the first professors at the Institute for Advanced Study in Princeton, New Jersey. Motivated by a continuing desire to develop mathematical techniques suited to quantum phenomena, from 1929 until the 1940s von Neumann introduced a theory of rings of operators, now known as von Neumann algebras. Other achievements include a proof of the quasi-ergodic hypothesis (1932) and important work in lattice theory (1935–7). It was not only the "new physics" that commanded von Neumann's attention. A 1932 Princeton lecture, *On Certain Equations of Economics and a Generalization of Brouwer's Fixed Point Theorem* (published 1937), was a seminal contribution to linear and nonlinear programming in economics.

Von Neumann became a legend in Princeton. It was said that he played practical jokes on Einstein, could recite verbatim books that he had read years earlier, and could edit assembly-language computer code in his head. Von Neumann's natural diplomacy helped him move easily among Princeton's intelligentsia, where he often adopted a tactful modesty. He once said he felt he had not lived up to all that had been expected of him. Never much like the stereotypical mathematician, he was known as a wit, bon vivant, and aggressive driver – his frequent auto accidents led to one Princeton intersection being dubbed "von Neumann corner".

In late 1943 von Neumann began work on the Manhattan Project (a US government programme developed to harness nuclear energy for military purposes) at the invitation of J. Robert Oppenheimer. Von Neumann was an expert in the nonlinear physics of hydrodynamics and shock waves, an expertise that he had already applied to chemical explosives in the British war effort. At Los Alamos, New Mexico, von Neumann worked on Seth Neddermeyer's implosion design for an atomic bomb. This called for a hollow sphere containing fissionable plutonium to be symmetrically imploded in order to drive the plutonium into a critical mass at the centre. Adapting an idea proposed by James Tuck, von Neumann calculated that a "lens" of faster- and slower-burning chemical explosives could achieve the requisite degree of symmetry. The "Fat Man" atomic bomb, dropped on the Japanese port of Nagasaki in 1945, used this design.

Overlapping with this work was von Neumann's magnum opus of applied mathematics, *Theory of Games and Economic Behavior* (1944), cowritten with the Princeton economist Oskar Morgenstern. Game theory had been orphaned since the 1928 publication of "Theory of Parlour Games", with neither von Neumann nor anyone else significantly developing it. The collaboration with Morgernstern burgeoned to 641 pages, the authors arguing for game theory as the "Newtonian science" underlying economic decisions. The book created a vogue for game theory among economists that has since partly subsided. The theory has also had broad influence in fields ranging from evolutionary biology to defence planning.

In the post-war years, von Neumann spent an increasing amount of time as a consultant to government and industry. Starting in 1944, he contributed important ideas for the U.S. Army's hard-wired ENIAC computer, designed by J. Presper Eckert Jr and John W. Mauchly. Most important, von Neumann modified the ENIAC to run as a stored-program machine. He then lobbied to build an improved computer at the Institute for Advanced Study. The IAS machine, which began operating in 1951, used binary arithmetic – the ENIAC had used decimal numbers – and shared the same memory for code and data, a design that greatly facilitated the "conditional loops" at the heart of all subsequent coding.

Another important consultancy was at the RAND Corporation, a think tank charged with planning nuclear strategy for the U.S. Air Force. Von Neumann insisted on the value of game-theoretic thinking in defence policy. He supported the development of the hydrogen bomb and was reported to have advocated a preventive nuclear strike to destroy the Soviet Union's nascent nuclear capability around 1950. Despite his hawkish stance, von Neumann defended Oppenheimer against attacks on his patriotism. From 1954 to 1956, von Neumann served as a member of the Atomic Energy Commission and was an architect of the policy of nuclear deterrence developed by President Dwight D. Eisenhower's administration.

Von Neumann was diagnosed with bone cancer in 1955, but continued to work even as his health deteriorated rapidly. In 1956 he received the Enrico Fermi Award. A lifelong agnostic, shortly before his death he converted to Roman Catholicism.

GEORGE GAMOW (1904–1968)

Russian-born American nuclear physicist
and cosmologist who was one of the foremost
advocates of the big bang theory.

Gamow was born in Odessa, in the Russian Empire (now in Ukraine). He attended Leningrad (now St. Petersburg) University, where he studied briefly with A.A. Friedmann, a mathematician and cosmologist who suggested that the universe should be expanding. At that time Gamow did not pursue Friedmann's suggestion, preferring instead to delve into quantum theory. After graduating in 1928, he travelled to Göttingen, where he developed his quantum theory of radioactivity – the first successful explanation of the behaviour of radioactive elements, some of which decay in seconds while others decay over thousands of years.

His achievement earned him a fellowship at the Copenhagen Institute of Theoretical Physics from 1928 to 1929, where he continued his investigations in theoretical nuclear physics. There he proposed his "liquid drop" model of atomic nuclei, which later served as the basis for the modern theories of nuclear fission and fusion. He also collaborated with F. Houtermans and R. Atkinson in developing a theory of the rates of thermonuclear reactions inside stars.

In 1934, after emigrating from the Soviet Union, Gamow was appointed professor of physics at George Washington University in Washington, D.C. There, in 1936, he collaborated with Edward Teller in developing a theory of beta decay – a nuclear decay process in which an electron is emitted.

Soon after, Gamow resumed his study of the relations between small-scale nuclear processes and cosmology. He used his knowledge of nuclear reactions to interpret stellar evolution, working with Teller on a theory of the internal structures of red giant stars (1942). From his work on stellar evolution, Gamow postulated that the sun's energy results from thermonuclear processes.

Gamow and Teller were both proponents of the expanding-universe theory that had been advanced by Friedmann, Edwin Hubble (1889–1953), and Georges Lemaître (1894–1966). Gamow, however, modified the theory, creating a version of it that became known as the big bang theory. He and Ralph Alpher published this theory in a paper called "The Origin of

Chemical Elements" (1948). This paper, attempting to explain the distribution of chemical elements throughout the universe, posits a primeval thermonuclear explosion, the "big bang" that began the universe. According to the theory, after the big bang, atomic nuclei were built up by the successive capture of neutrons by the initially formed simple atomic nuclei.

In 1954 Gamow's scientific interests grew to encompass biochemistry. He proposed the concept of a genetic code and maintained that the code was determined by the order of recurring triplets of nucleotides, the basic components of DNA. His proposal was vindicated during the rapid development of genetic theory that followed.

Gamow held the position of professor of physics at the University of Colorado, Boulder, from 1956 until his death. He is perhaps best known for his popular writings, designed to introduce to the nonspecialist such difficult subjects as relativity and cosmology. His first such work, *Mr. Tomkins in Wonderland* (1936), gave rise to the multivolume Mr. Tomkins series (1939–67). Among his other writings are *One, Two, Three . . . Infinity* (1947), *The Creation of the Universe* (1952), *A Planet Called Earth* (1963), and *A Star Called the Sun* (1964).

J. ROBERT OPPENHEIMER (1904–1967)

American theoretical physicist and science administrator, noted as director of the Los Alamos laboratory during development of the atomic bomb.

Oppenheimer was born in New York City. During his undergraduate studies at Harvard University, he excelled in Latin, Greek, physics, and chemistry; published poetry; and studied

Oriental philosophy. After graduating in 1925, he sailed for England to conduct research at the Cavendish Laboratory at the University of Cambridge. There Oppenheimer had the opportunity to collaborate with the British scientific community in its efforts to advance the cause of atomic research.

Max Born invited Oppenheimer to Göttingen University, where he met other prominent physicists, such as Niels Bohr and P.A.M. Dirac, and where, in 1927, he received his doctorate. After short visits to science centres in Leiden and Zürich, he returned to the United States to teach physics at the University of California at Berkeley and the California Institute of Technology.

In the 1920s the new quantum and relativity theories were engaging the attention of science. That mass was equivalent to energy and that matter could be both wavelike and corpuscular carried implications seen only dimly at that time. Oppenheimer's early research was devoted in particular to the energy processes of subatomic particles, including electrons, positrons, and cosmic rays. Since quantum theory had been proposed only a few years before, the university post provided him with an excellent opportunity to devote his entire career to the exploration and development of its full significance. In addition, he trained a whole generation of U.S. physicists, who were greatly influenced by his qualities of leadership and intellectual independence.

The rise of Adolf Hitler in Germany stirred Oppenheimer's first interest in politics. In 1936 he sided with the republic during the Civil War in Spain, where he became acquainted with Communist students. Although his father's death in 1937 left Oppenheimer a fortune that allowed him to subsidize anti-Fascist organizations, the tragic suffering inflicted by Joseph Stalin on Russian scientists led him to withdraw his associations with the Communist Party – in fact, he never joined the party – and at the same time reinforced in him a liberal democratic philosophy.

After the invasion of Poland by Nazi Germany in 1939, the physicists Albert Einstein and Leo Szilard warned the U.S. government of the danger threatening all of humanity if the Nazis should be the first to make a nuclear bomb. Oppenheimer then began to seek a process for the separation of uranium-235 from natural uranium and to determine the critical mass of uranium required to make such a bomb. In August 1942 the U.S. Army was given the responsibility of organizing the efforts of British and U.S. physicists to seek a way to harness nuclear energy for military purposes – an effort that became known as the Manhattan Project. Oppenheimer was instructed to establish and administer a laboratory to carry out this assignment. In 1943 he chose the plateau of Los Alamos, near Santa Fe, New Mexico.

The joint effort of outstanding scientists at Los Alamos culminated in the explosion of the first nuclear bomb on 16 July 1945, at the Trinity Site near Alamogordo, New Mexico, after the surrender of Germany. In October of the same year, Oppenheimer resigned his post. In 1947 he became head of the Institute for Advanced Study (IAS), Princeton, New Jersey, and served from 1947 until 1952 as chairman of the General Advisory Committee of the Atomic Energy Commission, which in October 1949 opposed development of the hydrogen bomb.

On 21 December 1953, he was notified of a military security report unfavourable to him and was accused of having associated with Communists in the past, of delaying the naming of Soviet agents, and of opposing the building of the hydrogen bomb. A security hearing declared him not guilty of treason but ruled that he should not have access to military secrets. As a result, his contract as adviser to the Atomic Energy Commission was cancelled.

The Federation of American Scientists immediately came to his defence with a protest against the trial. Oppenheimer was made the worldwide symbol of the scientist who, while trying to resolve the moral problems that arise from scientific discovery, becomes the victim of a witch-hunt. He spent the last years of his life working out ideas on the relationship between science and society. In 1963, President Lyndon B. Johnson presented Oppenheimer with the Enrico Fermi Award. Oppenheimer retired from the IAS in 1966 and died of throat cancer the following year.

KURT GÖDEL (1906–1978)

Austrian-born American mathematician, logician,
and author of Gödel's proof, a hallmark
of twentieth-century mathematics.

Gödel was born in Brünn, Austria-Hungary (now Brno, Czech Republic). He studied at the University of Vienna and received his doctorate in 1930. A member of the faculty of the university from 1930, Gödel was also a member of the Institute for Advanced Study, Princeton, New Jersey, between 1933 and 1952. He was a professor at the institute from 1953 until 1976, when he became professor emeritus. Gödel emigrated to the United States in 1940 and became a naturalized citizen in 1948. Gödel's proof first appeared in an article in the *Monatshefte für Mathematik und Physik* ("Monthly Magazine for Mathematics and Physics"), volume 38 (1931), on formally indeterminable propositions of the *Principia Mathematica* (1910–13) of Alfred North Whitehead and Bertrand Russell. Gödel's article ended nearly a century of attempts to establish axioms that would provide a rigorous basis for all mathematics, the most nearly (but, as Gödel showed, by no means entirely) successful attempt having been the *Principia Mathematica*. Another of Gödel's works is *Consistency of the Axiom of Choice and of the Generalized Continuum-Hypothesis with the Axioms of Set Theory* (1940), which became a classic of modern mathematics.

The ingenious argument in Gödel's proof was based on the observation that syntactical statements about the language of mathematics can be translated into statements of arithmetic, hence into the language of mathematics. It was partly inspired by an argument that supposedly goes back to the Ancient Greeks and may be paraphrased as: "Epimenides says that all Cretans are *liars*; Epimenides is a Cretan; hence Epimenides is a liar." Gödel showed that within any rigidly logical mathematical system there are propositions that cannot be proved or disproved on the basis of the axioms within that system. It is therefore uncertain that the basic axioms of arithmetic will not give rise to contradictions. This result had a profound effect on twentieth-century mathematics.

HANS BETHE (1906–2005)

German-born American theoretical physicist
who helped shape quantum physics.

Bethe was born in Strassburg, Germany (now Strasbourg, France). He started reading at the age of four and began writing at about the same age. His numerical and mathematical abilities also manifested themselves early. Bethe graduated from the gymnasium in the spring of 1924. After completing two years of study at the University of Frankfurt, he was advised by one of his teachers to go to the University of Munich and study with the physicist Arnold Sommerfeld. It was in Munich that Bethe discovered his exceptional proficiency in physics. He obtained a doctorate in 1928 with a thesis on electron diffraction in crystals. During 1930, as a Rockefeller Foundation fellow, Bethe spent a semester at the University of Cambridge under the aegis of Ralph Fowler and a semester at the University of Rome working with Enrico Fermi.

Bethe's craftsmanship combined the best of what he had learned from Sommerfeld and from Fermi: the thoroughness and rigor of the former and the clarity and simplicity of the latter. This skill was displayed in full force in the many reviews that Bethe wrote. His two book-length reviews in the 1933 *Handbuch der Physik* ("Handbook of Physics") – the first with Sommerfeld on solid-state physics and the second on the quantum theory of one- and two-electron systems – exhibited his remarkable powers of synthesis. Along with a review on nuclear physics in *Reviews of Modern Physics* (1936–7), these works were instant classics. All of Bethe's reviews were syntheses of the fields in question, giving them coherence and unity while charting the paths to be taken in addressing new problems. They usually contained much new material that Bethe had worked out in their preparation.

In the autumn of 1932, Bethe obtained an appointment at the University of Tübingen as an acting assistant professor of theoretical physics. In April 1933, after Adolf Hitler's accession to power, he was dismissed because his maternal grandparents were Jews. Sommerfeld was able to help him by awarding him a fellowship for the summer of 1933, and William Lawrence Bragg invited him to the University of Manchester for the following academic year. Bethe then went to the University of Bristol for the 1934 autumn semester before accepting a position at Cornell University, Ithaca, New York. He arrived at Cornell in February 1935 and stayed there for the rest of his life.

Bethe came to the United States at a time when the American physics community was undergoing enormous growth. As a result of what he learned at a conference on stellar energy generation in 1938, Bethe was able to give definitive answers to the problem of energy generation in stars. By stipulating and analysing the nuclear reactions responsible for the phenomenon, he explained how stars could continue to burn for billions of years. His 1939 *Physical Review* paper on energy generation in stars created the field of nuclear astrophysics and led to his being awarded the 1967 Nobel Prize for Physics.

During World War II Bethe first worked on problems in radar, spending a year at the Radiation Laboratory at the Massachusetts Institute of Technology. In 1943 he joined the Los Alamos Laboratory in New Mexico as the head of its theoretical division. He and the division were part of the Manhattan Project (a US government programme to harness nuclear energy for military purposes), and they made crucial contributions to the feasibility and design of the uranium and the plutonium atomic bombs. The years at Los Alamos changed Bethe's life.

In the aftermath of the development of these fission weapons, Bethe became deeply involved with investigating the feasibility of developing fusion bombs, hoping to prove that no terrestrial mechanism could accomplish the task. He believed their development to be immoral. When the Teller-Ulam mechanism for igniting a fusion reaction was advanced in 1951 and the possibility of a hydrogen bomb, or H-bomb, became a reality, Bethe helped to design it. He believed that the Soviets would likewise be able to build one and that only a balance of terror would prevent their use.

As a result of these activities, Bethe became deeply occupied with what he called "political physics" – the attempt to educate the public and politicians about the consequences of the existence of nuclear weapons. He became a relentless champion of nuclear arms control, writing many essays (collected in *The Road from Los Alamos*, 1991). He also became deeply committed to making peaceful applications of nuclear power economical and safe. Throughout his life, Bethe was a staunch advocate of nuclear power, defending it as an answer to the inevitable exhaustion of fossil fuels.

Bethe served on numerous advisory committees to the United States government, including the President's Science Advisory Committee. As a member of the committee, he helped persuade President Dwight D. Eisenhower to commit the United States to ban atmospheric nuclear tests. (The Nuclear Test Ban Treaty, which banned such tests, was finally ratified in 1963.) In 1972 Bethe's cogent and persuasive arguments helped prevent the deployment of antiballistic missile systems. He was influential in opposing President Ronald Reagan's Strategic Defense Initiative, arguing that a space-based laser defence system could be easily countered and that it would lead to further arms escalation. By virtue of these activities, and his general comportment, Bethe came to represent the science community's conscience. It was indicative of his constant grappling with moral issues that in 1995 he urged fellow scientists to collectively take a "Hippocratic oath" not to work on designing new nuclear weapons.

Throughout the political activism that marked his later life, Bethe never abandoned his scientific researches. Until well into his 90s he made important contributions at the frontiers of physics and astrophysics. He helped elucidate the properties of neutrinos and explained the observed rate of neutrino emission by the sun.

RACHEL CARSON (1907–1964)

American biologist well known for her
writings on environmental pollution
and the natural history of the sea.

Carson was born in Springdale, Pennsylvania. Early in her life
she developed a deep interest in the natural world. She entered
Pennsylvania College for Women with the intention of becoming a writer, but soon changed her major field of study
from English to biology. After taking her bachelor's degree in
1929, she continued her studies at Johns Hopkins University,
where she obtained her master's degree in 1932. In 1931 she
joined the faculty of the University of Maryland, where she
taught for five years. From 1929 to 1936 she also taught in the
Johns Hopkins summer school and pursued postgraduate
studies at the Marine Biological Laboratory in Woods Hole,
Massachusetts.

In 1936 Carson took a position as aquatic biologist with the
U.S. Bureau of Fisheries (from 1940 the U.S. Fish and Wildlife
Service), where she remained until 1952; for the last three
years as editor in chief of the service's publications. An article
in *The Atlantic Monthly* in 1937 served as the basis for her first
book, *Under the Sea-Wind*, published in 1941. It was widely
praised, as were all her books, for its remarkable combination
of scientific accuracy and thoroughness with an elegant and
lyrical prose style. *The Sea Around Us* (1951) became a
national best-seller, won a National Book Award, and was
eventually translated into 30 languages. Her third book, *The
Edge of the Sea*, was published in 1955.

Carson's prophetic *Silent Spring* (1962) was first serialized
in *The New Yorker* and then became a best-seller, creating
worldwide awareness of the dangers of environmental pollu-

tion. Carson stood behind her warnings of the consequences of indiscriminate pesticide use, despite the threat of lawsuits from the chemical industry and accusations that she engaged in "emotionalism" and "gross distortion". Carson died before she could see any substantive results from her work on this issue, but she left behind some of the most influential environmental writing ever published.

ALAN M. TURING (1912–1954)

British mathematician and logician who made major contributions to mathematics, cryptanalysis, logic, and to the new areas later named computer science, cognitive science, artificial intelligence, and artificial life.

Turing was born in London, the son of a British member of the Indian civil service. He entered King's College, University of Cambridge, to study mathematics in 1931. After graduating in 1934, he was elected to a fellowship at King's College in recognition of his research in probability theory. In 1936 Turing's seminal paper "On Computable Numbers, with an Application to the *Entscheidungsproblem* [Decision Problem]" was recommended for publication by the American mathematician-logician Alonzo Church, who had himself just published a paper that reached the same conclusion as Turing's. Later that year, Turing moved to Princeton University to study for a PhD in mathematical logic under Church's direction, which he completed in 1938.

The *Entscheidungsproblem* seeks an effective method for deciding which mathematical statements are provable within a given formal mathematical system and which are not. Turing and Church independently showed that in general this problem has no solution, proving that no consistent formal system of arithmetic is decidable. This result and others – notably work of the mathematician-logician Kurt Gödel (1906–78) – ended the dream of a system that could banish ignorance from mathematics forever. Turing introduced into his paper the idea that any effectively calculable function can be calculated by a universal Turing machine – a type of abstract computer that Turing had introduced in the course of his proof. In a review of Turing's work, Church acknowledged the superiority of Turing's formulation over his own, saying that the concept of computability by a Turing machine "has the advantage of making the identification with effectiveness . . . evident immediately."

In the summer of 1938 Turing returned from the United States to his fellowship at King's College. At the outbreak of hostilities with Germany in September 1939, he joined the wartime headquarters of the Government Code and Cypher School at Bletchley Park, Buckinghamshire. The British government had just been given the details of efforts by the Poles, assisted by the French, to break the Enigma code, used by the German military for their radio communications. As early as 1932, a small team of Polish mathematician-cryptanalysts, led by Marian Rejewski, had succeeded in reconstructing the internal wiring of the type of Enigma machine used by the Germans, and by 1938 they had devised a code-breaking machine, code-named "Bomba" (the Polish word for a type of ice cream).

The Bomba depended for its success on German operating procedures, and a change in procedures in May 1940 rendered it virtually useless. During 1939 and the spring of 1940, Turing and others designed a radically different code-breaking machine known as the Bombe. Turing's ingenious Bombes kept the Allies supplied with intelligence for the remainder of the war. By early 1942 the Bletchley Park cryptanalysts were decoding about 39,000 intercepted messages each month, which subsequently rose to more than 84,000 per month. At the end of the war, Turing was made an officer of the Order of the British Empire for his code-breaking work.

In 1945, the war being over, Turing was recruited to the National Physical Laboratory (NPL) in London to design and develop an electronic computer. His design for the Automatic Computing Engine (ACE) was the first relatively complete specification of an electronic stored-program general-purpose digital computer. Had Turing's ACE been built as planned, it would have had considerably more memory than any of the other early computers, as well as being faster. However, his colleagues at NPL thought the engineering too difficult to attempt, and a much simpler machine was built, the Pilot Model ACE.

In the end, NPL lost the race to build the world's first working electronic stored-program digital computer – an honour that went to the Royal Society Computing Machine Laboratory at the University of Manchester in June 1948. Discouraged by the delays at NPL, Turing took up the deputy directorship of the Computing Machine Laboratory in that year (there was no director). His earlier theoretical concept of a universal Turing machine had been a fundamental influence on the Manchester computer project from its inception. Turing's principal practical contribution after his arrival at Manchester was to design the programming system of the Ferranti Mark I, the world's first commercially available electronic digital computer.

Turing was a founding father of modern cognitive science and a leading early exponent of the hypothesis that the human brain is in large part a digital computing machine. He theorized that the cortex at birth is an "unorganised machine" that through "training" becomes organized "into a universal machine or something like it." A pioneer of artificial intelligence, Turing proposed in 1950 what subsequently became known as the Turing test as a criterion for whether a machine thinks. Such a test consists of a remote human interrogator, within a fixed time frame, attempting to distinguish between a computer and a human subject based on their replies to various questions posed by the interrogator.

Though he was elected a fellow of the Royal Society in March 1951, Turing's life was about to suffer a major setback. In March 1952 he was prosecuted for homosexuality, then a crime in Britain, and sentenced to 12 months of hormone "therapy" – a treatment that he seems to have borne with amused fortitude. Judged a security risk by the British government, Turing lost his security clearance and his access to ongoing government work with codes and computers. He spent the rest of his short career at the University of Manchester, where he was appointed to a specially created readership in the theory of computing in May 1953.

From 1951 Turing had been working on what is now known as artificial life. He wrote "The Chemical Basis of Morphogenesis", which described some of his research on the development of pattern and form in living organisms, and he used the Ferranti Mark I computer to model chemical mechanisms by which genes could control the development of anatomical structure in plants and animals.

NORMAN ERNEST BORLAUG (1914–)

American agricultural scientist and plant pathologist.

Borlaug was born in Cresco, Iowa. He studied plant biology and forestry at the University of Minnesota and earned a PhD in plant pathology there in 1941. From 1944 to 1960 he served as research scientist at the Rockefeller Foundation's Cooperative Mexican Agricultural Program in Mexico. At a research station at Campo Atizapan he developed strains of grain that dramatically increased crop yields. Wheat production in Mexico multiplied threefold in the time that he worked with the Mexican government; "dwarf" wheat imported in the mid-1960s was responsible for a 60-per-cent increase in harvests in Pakistan and India. He also created a wheat–rye hybrid known as triticale. The increased yields resulting from Borlaug's new strains enabled many developing countries to become agriculturally self-sufficient. For this work Borlaug was awarded the Nobel Prize for Peace in 1970.

Borlaug served as director of the Inter-American Food Crop Program from 1960 to 1963 and as director of the International Maize and Wheat Improvement Center, Mexico City, from 1964 to 1979. In constant demand as a consultant, he has served on numerous committees and advisory panels on agriculture, population control, and renewable resources.

SIR FRED HOYLE (1915–2001)

British mathematician and astronomer, best known
as the foremost proponent and defender of
the steady state theory of the universe.

Hoyle was born in Bingley, Yorkshire. He was educated at
Emmanuel College and St John's College, University of Cambridge. During World War II he worked with the British
Admiralty on radar development, and in 1945 he returned
to Cambridge as a lecturer in mathematics. Three years later,
in collaboration with the astronomer Thomas Gold and the
mathematician Hermann Bondi, he announced their steady
state theory. (This theory holds that the universe is always
expanding but maintaining a constant average density; matter
being continuously created to form new stars and galaxies at
the same rate that old ones become unobservable as a result of
their increasing distance and velocity. A steady-state universe
has no beginning or end in time.) Within the framework of
Albert Einstein's theory of relativity, Hoyle formulated a
mathematical basis for the steady-state theory, making the
expansion of the universe and the creation of matter interdependent.

In the late 1950s and early '60s, controversy about the
steady state theory grew. New observations of distant galaxies
and other phenomena, supporting the big bang theory, weakened the steady state theory, and it subsequently fell out of
favour with most cosmologists. Although Hoyle was forced to
alter some of his conclusions, he tenaciously tried to make his
theory consistent with new evidence.

Hoyle was elected to the Royal Society in 1957, a year after
joining the staff of the Hale Observatories (now the Mount
Wilson and Palomar observatories). In collaboration with

William Fowler and others in the United States, he formulated theories about the origins of stars as well as about the origins of elements within stars. Hoyle was director of the Institute of Theoretical Astronomy at Cambridge (1967–73), an institution he was instrumental in founding. He received a knighthood in 1972.

Hoyle is known for his popular science works, including *The Nature of the Universe* (1951), *Astronomy and Cosmology* (1975), and *The Origin of the Universe and the Origin of Religion* (1993). He also wrote novels, plays, short stories, and an autobiography, *The Small World of Fred Hoyle* (1986).

FRANCIS HARRY COMPTON CRICK (1916–2004) AND JAMES DEWEY WATSON (1928–)

Biophysicists who uncovered the molecular structure of deoxyribonucleic acid (DNA), the chemical substance ultimately responsible for hereditary control of life functions.

Crick was born in Northampton, England. During World War II he interrupted his education to work as a physicist in the development of magnetic mines for use in naval warfare, but afterwards turned to biology at the Strangeways Research Laboratory, University of Cambridge (1947). Interested in pioneering efforts to determine the three-dimensional structures of large molecules found in living organisms, he transferred to the university's Medical Research Council Unit at the Cavendish Laboratory in 1949.

In 1951, the American biologist James Watson arrived at the

laboratory. Watson, born in Chicago, Illinois, had enrolled at the University of Chicago when only 15 and graduated in 1947. From his virus research at Indiana University, where he obtained his PhD in 1950, and from the experiments of the Canadian-born American bacteriologist Oswald Avery, which proved that DNA affects hereditary traits, Watson became convinced that the gene could be understood only after something was known about nucleic acid molecules. He learned that scientists working in the Cavendish Laboratory at Cambridge were using photographic patterns made by X-rays that had been shot through protein crystals to study the structure of protein molecules.

It was known that the mysterious nucleic acids, especially DNA, played a central role in the hereditary determination of the structure and function of each cell. Watson convinced Crick that knowledge of DNA's three-dimensional structure would make its hereditary role apparent. Using the X-ray diffraction studies of DNA done by Maurice Wilkins and X-ray diffraction pictures produced by Rosalind Franklin, Watson and Crick were able to construct a molecular model consistent with the known physical and chemical properties of DNA. The model consisted of two intertwined helical (spiral) strands of sugar-phosphate, bridged horizontally by flat organic bases. Watson and Crick theorized that if the strands of the double helix were separated, each would serve as a template for the formation, from small molecules in the cell, of a new sister strand identical to its former partner. This copying process explained the replication of the gene and, eventually, the chromosome, that was known to occur in dividing cells. Watson and Crick's model also indicated that the sequence of bases along the DNA molecule spells some kind of code, which is "read" by a cellular mechanism that translates it into the specific proteins responsible for a cell's particular structure and function.

Watson and Crick published their epochal discovery in two papers in 1953. The research answered one of the fundamental questions in genetics, and in 1962 Crick, Watson, and Wilkins were awarded the Nobel Prize for Physiology and Medicine. By 1961 Crick had evidence to show that each group of three bases (a codon) on a single DNA strand designates the position of a specific amino acid on the backbone of a protein molecule. He also helped to determine which codons code for each of the 20 amino acids normally found in protein, and thus helped clarify the way in which the cell eventually uses the DNA "message" to build proteins.

From 1977 until his death, Crick held the position of distinguished professor at the Salk Institute for Biological Studies in San Diego, California, where he conducted research on the neurological basis of consciousness. His book *Of Molecules and Men* (1966) discusses the implications of the revolution in molecular biology. *What Mad Pursuit: A Personal View of Scientific Discovery* was published in 1988. In 1991 Crick received the Order of Merit.

Following his collaboration with Crick, Watson taught at Harvard University from 1955 to 1976, where he served as professor of biology from 1961 to 1976. He conducted research on the role of nucleic acids in the synthesis of proteins. In 1965 he published *Molecular Biology of the Gene*, one of the most extensively used modern biology texts. He later wrote *The Double Helix* (1968), an informal and personal account of the DNA discovery and the roles of the people involved in it, which aroused some controversy. In 1968 Watson assumed the leadership of the Laboratory of Quantitative Biology at Cold Spring Harbor, Long Island, New York, and made it a world centre for research in molecular biology. He concentrated its efforts on cancer research. In 1981 *The DNA Story*, co-authored with John Tooze, was published.

From 1988 to 1992 at the U.S. National Institutes of Health, Watson helped direct the Human Genome Project – a project to map and decipher all the genes in the human chromosomes – but he eventually resigned because of alleged conflicts of interest involving his investments in private biotechnology companies. In early 2007 Watson's own genome was sequenced and made publicly available on the Internet. He was the second person in history to have a personal genome sequenced in its entirety. In October of the same year, he sparked controversy by making a public statement alluding to the idea that the intelligence of Africans might not be the same as that of other peoples, and that intellectual differences among geographically separated peoples might arise over time as a result of genetic divergence. Watson's remarks were immediately denounced as racist. Though he denied this charge, he resigned from his position at Cold Spring Harbor and formally announced his retirement less than two weeks later.

RICHARD P. FEYNMAN (1918–1988)

American theoretical physicist who was widely regarded as the most brilliant, influential, and iconoclastic figure in his field in the post-World-War-II era.

Born in the Far Rockaway section of New York City, Feynman was the descendant of Russian and Polish Jews who had immigrated to the United States in the late nineteenth century. He studied physics at the Massachusetts Institute of Technology, where his undergraduate thesis (1939) proposed an original and enduring approach to calculating forces in molecules. Feynman received his doctorate at Princeton University in 1942. At Princeton, with his adviser, John Archibald Wheeler, he developed an approach to quantum mechanics governed by the principle of least action. This approach replaced the wave-oriented electromagnetic picture developed by James Clerk Maxwell (1831–79) with one based entirely on particle interactions mapped in space and time. In effect, Feynman's method calculated the probabilities of all the possible paths a particle could take in going from one point to another.

During World War II Feynman was recruited to serve as a staff member of the U.S. atomic bomb project at Princeton University in 1941–2, and then at the new secret laboratory at Los Alamos, New Mexico, from 1943 to 1945. At Los Alamos he became the youngest group leader in the theoretical division of the Manhattan Project (a US government programme to harness nuclear energy for military purposes). With the head of that division, Hans Bethe, he devised the formula for predicting the energy yield of a nuclear explosive.

Feynman also took charge of the project's primitive computing effort, using a hybrid of new calculating machines and human workers to try to process the vast amounts of numerical computation required by the project. He observed the first detonation of an atomic bomb on 16 July 1945, near Alamogordo, New Mexico, and, although his initial reaction was euphoric, he later felt anxiety about the force he and his colleagues had helped unleash on the world.

From 1945 to 1950 Feynman served as an associate professor at Cornell University, and returned to studying the fundamental issues of quantum electrodynamics. In the years that followed, his vision of particle interaction kept returning to the forefront of physics as scientists explored esoteric new domains at the subatomic level. In 1950 he became professor of theoretical physics at the California Institute of Technology (Caltech), where he remained for the rest of his career.

Five particular achievements of Feynman stand out as crucial to the development of modern physics. First, and most important, is his work in correcting the inaccuracies of earlier formulations of quantum electrodynamics – the theory that explains the interactions between electromagnetic radiation (photons) and charged subatomic particles such as electrons and positrons (antielectrons). By 1948 Feynman completed this reconstruction of a large part of quantum mechanics and electrodynamics and resolved the meaningless results that the old quantum electrodynamic theory sometimes produced. He was co-awarded the Nobel Prize for Physics in 1965 for this work, which tied together in an experimentally perfect package all the varied phenomena at work in light, radio, electricity, and magnetism. The other co-winners of the prize, Julian S. Schwinger of the United States and Tomonaga Shin'ichiro of Japan, had independently created equivalent theories, but it was Feynman's that proved the most original and far-reaching.

Second, Feynman introduced simple diagrams, now called Feynman diagrams, that are easily visualized graphic analogues of the complicated mathematical expressions needed to describe the behaviour of systems of interacting particles. This work greatly simplified some of the calculations used to observe and predict such interactions.

Third, in the early 1950s Feynman provided a quantum-mechanical explanation for the Soviet physicist Lev D. Landau's theory of superfluidity – i.e. the strange, frictionless behaviour of liquid helium at temperatures near absolute zero. Fourth, in 1958 he and the American physicist Murray Gell-Mann devised a theory that accounted for most of the phenomena associated with the weak force, which is the force at work in radioactive decay. Their theory, which turns on the asymmetrical "handedness" of particle spin, proved particularly fruitful in modern particle physics. And finally, in 1968, while working with experimenters at the Stanford Linear Accelerator on the scattering of high-energy electrons by protons, Feynman invented a theory of "partons", or hypothetical hard particles inside the nucleus of the atom, that helped lead to the modern understanding of quarks.

Feynman's stature among physicists transcended even his sizable contributions to the field. His bold and colourful personality, unencumbered by false dignity or notions of excessive self-importance, seemed to announce: "Here is an unconventional mind." He was a master calculator who could create a dramatic impression in a group of scientists by slashing through a difficult numerical problem. His fellow physicists envied his flashes of inspiration and admired him for other qualities too: a faith in nature's simple truths, a scepticism about official wisdom, and an impatience with mediocrity.

Feynman's lectures at Caltech evolved into the books *Quantum Electrodynamics* (1961) and *The Theory of Fundamental Processes* (1961). In 1961 he began reorganizing and teaching the introductory physics course at Caltech; the result, published as *The Feynman Lectures on Physics* (1963–5), became a classic textbook. Feynman's views on quantum mechanics, scientific method, the relationship between science and religion, and the role of beauty and uncertainty in scientific knowledge are expressed in two models of science writing, again distilled from lectures: *The Character of Physical Law* (1965) and *QED: The Strange Theory of Light and Matter* (1985).

ROSALIND FRANKLIN (1920–1958)

British scientist who contributed to the discovery of the molecular structure of deoxyribonucleic acid (DNA).

Franklin was born in London. She studied physical chemistry at Newnham College, University of Cambridge, graduating in 1941. She then joined the British Coal Utilisation Research Association, where she contributed to studies that explained the absorption properties of coals. From 1947 to 1950 she worked with Jacques Méring at the State Chemical Laboratory in Paris, studying X-ray diffraction technology. That work led to her research on the structural changes caused by the formation of graphite in heated carbons – work that proved valuable for the coking industry.

In 1951 Franklin joined the Biophysical Laboratory at King's College, London, where she applied X-ray diffraction methods to the study of DNA. She is credited with discoveries that established the density of DNA, its helical conformation, and other significant aspects.

From 1953 to 1958 Franklin worked in the Crystallography Laboratory at Birkbeck College, London. While there she completed her work on coals and on DNA and began a project on the molecular structure of the tobacco mosaic virus. She collaborated on studies showing that the ribonucleic acid (RNA) in that virus was embedded in its protein rather than in its central cavity, and that this RNA was a single-strand helix, rather than the double helix found in the DNA of bacterial viruses and higher organisms.

JACK KILBY (1923–2005)

American engineer and one of the inventors of the integrated circuit, a system of interconnected transistors on a single microchip.

Born in Jefferson City, Missouri, Kilby was the son of an electrical engineer. Like many inventors of his era, he gained his start in electronics with amateur radio. His interest began while he was in high school when the Kansas Power Company of Great Bend, Kansas, of which his father was president, had to rely on amateur radio operators for communications after an ice storm disrupted normal service. After serving as an electronics technician in the U.S. Army during World War II, Kilby enrolled in the electrical engineering programme at the University of Illinois in Urbana-Champaign, gaining his bachelor's degree in 1947.

After graduation Kilby joined the Centralab Division of Globe Union Incorporated, located in Milwaukee, Wisconsin, where he was placed in charge of designing and developing miniaturized electronic circuits. He also found time to continue his studies at the University of Wisconsin, Milwaukee Extension Division, and received his master's degree in 1950. In 1952 Centralab sent Kilby to Bell Laboratories' headquarters in Murray Hill, New Jersey, to learn about the transistor, which had been invented at Bell in 1947 and which Centralab had purchased a licence to manufacture. Back at Centralab, Kilby began working on germanium-based transistors for use in hearing aids. He soon realized, however, that he needed the resources of a larger company to pursue the goal of miniaturizing circuits, and in 1958 he switched to another Bell licensee, Texas Instruments Incorporated of Dallas, Texas.

Shortly after his arrival at Texas Instruments (TI), Kilby had his epoch-making "monolithic idea". He realized that, instead of connecting separate components, an entire electronic assembly could be made as one unit from one semiconducting material by overlaying it with various impurities to replicate individual electronic components, such as resistors, capacitors, and transistors. Soon Kilby had a working postage-stamp-sized prototype manufactured from germanium, and in February 1959 TI filed a patent application for this "miniaturized electronic circuit" – the world's first integrated circuit (IC).

Four months later, Robert Noyce of Fairchild Semiconductor Corporation filed a patent application for essentially the same device, but based on a different manufacturing procedure. Ten years later, long after their respective companies had cross-licensed technologies, the courts gave Kilby credit for the idea of the integrated circuit but gave Noyce the patent for his planar manufacturing process, a method for evaporating lines of conductive metal (the "wires") directly on to a silicon chip.

Although the original integrated circuit was Kilby's most important invention, it was only one of more than 50 patents that he was awarded. Many of those patents concerned improvements in IC design and manufacturing. In 1967 he designed the first IC-based electronic calculator, the Pocketronic, gaining himself and TI the basic patent that lies at the heart of all pocket calculators. The Pocketronic required dozens of ICs, making it too complicated and expensive to manufacture for consumers, but by 1972 TI had reduced the number of necessary ICs to one. Kilby began a leave of absence from TI in 1970 to pursue independent research, particularly in solar power generation, although he continued as a semiconductor consultant on a part-time basis. Among his many honours, Kilby was awarded the National Medal of Science in 1970, the Charles Stark Draper Medal in 1989, and the National Medal of Technology in 1990. In 1997 TI dedicated to him its new research and development building in Dallas, the Kilby Center. The Royal Swedish Academy of Sciences, breaking with a trend of recognizing only theoretical physicists, awarded half of the 2000 Nobel Prize for Physics to Kilby for his work as an applied physicist.

JOHN FORBES NASH JR (1928–)

American mathematician noted for his landmark work on the mathematics of game theory.

Nash was born in Bluefield, West Virginia. In 1948 he received bachelor's and master's degrees in mathematics from the Carnegie Institute of Technology (now Carnegie-Mellon University) in Pittsburgh, Pennsylvania. Two years later, at the age of 22, he completed his doctorate at Princeton University, publishing his influential thesis "Non-cooperative Games" in the journal *Annals of Mathematics*. He joined the faculty of the Massachusetts Institute of Technology in 1951 but resigned in the late 1950s after bouts of mental illness. He then began an informal association with Princeton.

Nash established the mathematical principles of game theory, a branch of mathematics that examines the rivalries among competitors with mixed interests. Known as the Nash solution or the Nash equilibrium, his theory attempted to explain the dynamics of threat and action among competitors. Despite its practical limitations, the Nash solution was widely applied by business strategists.

Nash showed that given a game with a set of possible outcomes and associated utilities (preferences) for each player, there is a unique outcome that satisfies four conditions. (1) The outcome is independent of the choice of a utility function (that is, if a player prefers x to y, the solution will not change if one function assigns to x a utility of 10 and to y a utility of 1 or a second function assigns the values of 20 and 2). (2) Both players cannot do better simultaneously (a condition known as Pareto-optimality). (3) The outcome is independent of irrelevant alternatives (in other words, if unattractive options are added to or dropped from the list of alternatives, the solution will not change). (4) The outcome is symmetrical (that is, if the players reverse their roles, the solution will remain the same, except that the payoffs will be reversed). In some cases the Nash solution seems inequitable because it is based on a balance of threats – the possibility that no agreement will be reached, so that both players will suffer losses – rather than a "fair" outcome.

For his work in game theory Nash was awarded the 1994 Nobel Prize for Economics. A film version of Nash's life, *A Beautiful Mind* (2001), based on Sylvia Nasar's 1998 biography of the same name, won an Academy Award for best picture. It portrays Nash's long struggle with schizophrenia.

EDWARD O. WILSON (1929–)

American biologist recognized as the world's leading authority on ants and the foremost proponent of sociobiology.

Wilson was born in Birmingham, Alabama. He gained his bachelor's and master's degrees in 1949 and 1950 at the

University of Alabama, and devoted much of his early career there to the study of ants. In the same year that he gained his doctorate (1955) at Harvard University, he completed an exhaustive taxonomic analysis of the ant genus *Lasius*. In collaboration with W.L. Brown, he developed the concept of "character displacement", a process in which two closely-related species populations undergo rapid evolutionary differentiation after first coming into contact with each other, in order to minimize the chances of both competition and hybridization between them.

After his appointment to the Harvard faculty in 1956, Wilson made a series of important discoveries, including the determination that ants communicate primarily through the transmission of a chemical substance known as a pheromone. In the course of revising the classification of ants in the South Pacific, he formulated the concept of the taxon cycle, in which speciation and species dispersal are linked to the varying habitats that organisms encounter as their populations expand.

In 1971 Wilson published *The Insect Societies*, his definitive work on ants and other social insects. The book provides a comprehensive picture, examining the ecology and population dynamics of innumerable species in addition to their societal behaviour patterns. In his second major work, *Sociobiology: The New Synthesis* (1975), Wilson presented his theories about the biological basis of social behaviour. One of the central tenets of sociobiology is that genes (and their transmission through successful reproduction) are the central motivators in animals' struggle for survival, and that animals will behave in ways that maximize their chances of transmitting copies of their genes to succeeding generations. Since behaviour patterns are to some extent inherited, the evolutionary process of natural selection can be said to foster those behavioural (as well as physical) traits that increase an individual's chances of reproducing.

Sociobiology has contributed several insights to the understanding of animal social behaviour. It explains apparently altruistic behaviour in some animal species as actually being genetically selfish, since such behaviours usually benefit closely related individuals whose genes resemble those of the altruistic individual. This insight helps explain why soldier ants sacrifice their lives in order to defend their colony, or why worker honeybees in a hive forego reproduction in order to help their queen reproduce. Sociobiology can in some cases explain the differences between male and female behaviour in certain animal species as resulting from the different strategies the sexes must resort to in order to transmit their genes to posterity.

One chapter in *Sociobiology* proposed that the essentially biological principles on which animal societies were based applied to human social behaviour. This inflamed certain scientists and groups, which regarded such ideas as politically provocative. Actually, Wilson maintained that he saw perhaps as little as 10 per cent of human behaviour as genetically induced; the rest being attributable to environment. In his 1979 Pulitzer Prize-winning book *On Human Nature* (1978), Wilson explored the implications of sociobiology with regard to human aggression, sexuality, and ethics. His book *The Ants* (1990) was a monumental summary of contemporary knowledge of those insects. In *The Diversity of Life* (1992), Wilson traced how the world's living species became diverse and examined the massive species extinctions caused by human activities in the twentieth century. His autobiography, *Naturalist*, appeared in 1994.

At Harvard, Wilson was professor of zoology from 1964 to 1976, and Frank B. Baird, Jr, Professor of Science thereafter. He was also curator of entomology at the Museum of Comparative Zoology from 1972. In 1990 he shared Sweden's Crafoord Prize with the American biologist Paul Ehrlich.

JANE GOODALL (1934–)

British ethologist, known for her exceptionally
detailed and long-term research on the chimpanzees
of Gombe Stream National Park in Tanzania.

Goodall was born in London. Interested in animal behaviour
from an early age, she left school at the age of 18 and worked
as a secretary and as a film production assistant until she
gained passage to Africa. Once there, Goodall began assisting
paleontologist and anthropologist Louis Leakey. Her association with Leakey led eventually to her establishment in June
1960 of a camp in the Gombe Stream Game Reserve (now a
national park) so that she could observe the behaviour of
chimpanzees in the region.

The University of Cambridge in 1965 awarded Goodall a
PhD in ethology; she was one of very few candidates to receive
a PhD without having first possessed a BA degree. Except for
short periods of absence, Goodall and her family remained in
Gombe until 1975, often directing the fieldwork of other
doctoral candidates. In 1977 she co-founded the Jane Goodall
Institute for Wildlife Research, Education, and Conservation
in California; the centre later moved its headquarters to
Washington, D.C.

Over the years Goodall was able to correct a number of
misunderstandings about chimpanzees. She found, for example, that the animals are omnivorous, not vegetarian; that they
are capable of making and using tools; and, in short, that they
have a set of hitherto unrecognized complex and highly
developed social behaviours. Goodall wrote a number of
books and articles about various aspects of her work, notably
In the Shadow of Man (1971). She summarized her years of
observation in *The Chimpanzees of Gombe: Patterns of Be-*

havior (1986). Goodall continued to write and lecture about environmental and conservation issues into the early twenty-first century. The recipient of numerous honours, she was created Dame of the British Empire in 2003.

SIR HAROLD W. KROTO (1939–), RICHARD E. SMALLEY (1943–2005), AND ROBERT F. CURL JR (1933–)

Chemists who together discovered
the carbon compounds called fullerenes.

Kroto was born in Wisbech, Cambridgeshire, England. He received a PhD from the University of Sheffield in 1964, then joined the faculty of the University of Sussex in 1967 and became a professor of chemistry there in 1985. In the course of his research, Kroto used microwave spectroscopy to discover long, chainlike carbon molecules in the atmospheres of stars and gas clouds. Wishing to study the vaporization of carbon in order to find out how these carbon chains formed, he went to Rice University, Houston, Texas, where he met with Curl and Smalley, who had designed an instrument – the laser-super-sonic cluster beam apparatus – that could vaporize almost any known material and then be used to study the resulting clusters of atoms or molecules.

Smalley, who was born in Akron, Ohio, was a leading proponent of the development and application of nanotech-nology – the manipulation of materials at the extremely small scale of individual atoms or groups of atoms. He joined the faculty of Rice University in 1976 after receiving his PhD in chemistry in 1973 from Princeton University. Curl, who was born in Alice, Texas, completed his doctoral studies in chemistry at the University of California at Berkeley in 1957 and joined the faculty at Rice in 1958.

In a series of experiments conducted in September 1985 at Rice University, the scientists simulated the chemistry in the atmosphere of giant stars by turning the vaporization laser on to graphite. The study not only confirmed that carbon chains were produced but also showed, serendipitously, that a hitherto unknown carbon species containing 60 atoms formed spontaneously in relatively high abundance. Attempts to explain the remarkable stability of the C_{60} cluster led the three to the conclusion that the cluster must be a spheroidal closed cage in the form of a truncated icosahedron (a polygon with 60 vertices and 32 faces, 12 of which are pentagons and 20 hexagons), which resembled a hollow sphere or ball. They chose the imaginative name buckminsterfullerene for the cluster, in honour of the designer-inventor of the geodesic domes whose ideas had influenced their structure conjecture.

Prior to the discovery of fullerenes – these cage-like molecules of carbon – only two well-defined allotropes (forms) of carbon were known: diamond (composed of a three-dimensional crystalline array of carbon atoms) and graphite (composed of stacked sheets of two-dimensional hexagonal arrays of carbon atoms). The fullerenes constitute a third form, and it is remarkable that their existence evaded discovery until almost the end of the twentieth century. For their discovery Kroto, Smalley, and Curl were awarded the 1996 Nobel Prize for Chemistry.

Other fullerenes were soon discovered. Smalley's later research focused on long cylindrical fullerenes called carbon nanotubes, which are extremely strong and have useful electrical properties. Kroto, working with colleagues at the University of Sussex, used laboratory microwave spectroscopy techniques to analyse the spectra of carbon chains. These measurements later led to the detection, by radioastronomy, of chain-like molecules consisting of five to eleven carbon atoms in interstellar gas clouds and in the atmospheres of carbon-rich red giant stars.

STEPHEN JAY GOULD (1941–2002)

American paleontologist, evolutionary biologist, and science writer.

Gould was born in New York City. He graduated from Antioch College in 1963 and received a PhD in paleontology at Columbia University in 1967. He joined the faculty of Harvard University in 1967, becoming a full professor there in 1973. Gould's own technical research focused on the evolution and speciation of West Indian land snails. With Niles Eldredge, he developed in 1972 the theory of punctuated equilibrium – a revision of Darwinian theory proposing that the creation of new species through evolutionary change occurs not at slow, constant rates over millions of years but rather in rapid bursts over periods as short as thousands of years, which are then followed by long periods of stability during which organisms undergo little further change. Gould's theory, as well as much of his later work, drew criticism from a number of other scientists.

Apart from his technical research, Gould became widely known as a writer, polemicist, and popularizer of evolutionary theory. In this capacity he was often engaged in defending evolution against the attacks of proponents of biblical creationism, who held that living things were created by God out of nothing.

Gould's science writing is characterized by a graceful literary style and the ability to treat complex concepts with absolute clarity. Among his diverse works are an exploration of the relationship between evolution and the development of individual organisms, *Ontogeny and Phylogeny* (1977); a discussion of intelligence testing and a refutation of claims for the intellectual superiority of some races, *The Mismeasure of Man* (1981), which won the National Book Critics Circle Award in 1982; and what was considered his magnum opus, the 1,433-page summary of his life's work, *The Structure of Evolutionary Theory* (2002). His volumes of collected Natural History essays include *Ever Since Darwin* (1977); *The Panda's Thumb* (1980), for which he received the National Book Award in 1981; *Hen's Teeth and Horse's Toes* (1983); and *I Have Landed: The End of a Beginning in Natural History* (2002), which was published the day after his death. Gould was the recipient of numerous honours: he received a MacArthur fellowship in 1981, the first year that grant was awarded; he became a member of the American Academy of Arts and Sciences in 1983 and the National Academy of Sciences in 1989; and he served as president of such organizations as the Paleontological Society (1985–6), the Society for the Study of Evolution (1990–1), and the American Association for the Advancement of Science (1999–2000).

STEPHEN W. HAWKING (1942–)

English theoretical physicist whose theory of
exploding black holes drew upon both relativity
theory and quantum mechanics.

Hawking was born in Oxford. He studied mathematics and physics at University College, University of Oxford, gaining his bachelor's degree in 1962; and Trinity Hall, University of Cambridge, where he achieved his PhD in 1966. He was elected a research fellow at Gonville and Caius College at Cambridge. In the early 1960s Hawking contracted amyotrophic lateral sclerosis, an incurable degenerative neuromuscular disease. He continued to work despite the disease's progressively disabling effects.

Hawking worked primarily in the field of general relativity and particularly on the physics of black holes. In 1971 he suggested that numerous tiny primordial black holes, possibly with a mass equal to that of an asteroid or less, might have been created during the big bang (a state of extremely high temperatures and density in which the universe is thought to have originated roughly 15 billion years ago). These objects, called mini black holes, are unique in that their immense mass and gravity require that they be ruled by the laws of relativity, while their minute size requires that the laws of quantum mechanics also apply to them. In 1974 Hawking proposed that, in accordance with the predictions of quantum theory, black holes emit subatomic particles. If a proton and an antiproton escape the black hole's gravitational attraction, they annihilate each other and in so doing generate energy – energy that they in effect drain from the black hole. If this process is repeated again and again, the black hole evaporates, having lost all of its energy and thereby its mass, since these are equivalent.

Hawking's work greatly spurred efforts to theoretically delineate the properties of black holes, objects about which it was previously thought that nothing could be known. His work was also important because it showed these properties' relationship to the laws of classical thermodynamics and quantum mechanics.

Hawking's contributions to physics have earned him many exceptional honours. In 1974 the Royal Society elected him one of its youngest fellows. He became professor of gravitational physics at Cambridge in 1977, and in 1979 he was appointed to Cambridge's Lucasian professorship of mathematics, a post once held by Isaac Newton. His publications include *The Large Scale Structure of Space-Time* (1973; co-authored with G.F.R. Ellis), *Superspace and Supergravity* (1981), *The Very Early Universe* (1983), and the best-seller *A Brief History of Time: From the Big Bang to Black Holes* (1988).

J. CRAIG VENTER (1946–) AND FRANCIS COLLINS (1950–)

Two American scientists who led projects
that mapped the human genome.

Venter was born in Salt Lake City, Utah. Soon after, his family moved to the San Francisco area, where swimming and surfing occupied his free time. After high school Venter joined the U.S. Naval Medical Corps and served in Vietnam. On returning to the States, he earned a BA in biochemistry in 1972, then a doctorate in physiology and pharmacology in 1975 at the University of California, San Diego. In 1976 he joined the faculty of the State University of New York at Buffalo, where he was involved in neurochemistry research. In 1984 Venter moved to the National Institutes of Health (NIH), studying genes involved in signal transmission between nerve cells.

While at the NIH, Venter became frustrated with traditional methods of gene identification, which were slow and time-consuming. He developed an alternative technique that he used to identify thousands of human genes much more quickly. Although first received with scepticism, the approach later gained increased acceptance, and in 1993 it was used to identify the gene responsible for a type of colon cancer. Venter's attempts to patent the gene fragments that he identified, however, created a furore among those in the scientific community who believed that such information belonged in the public domain.

Venter left the NIH in 1992 and, with the backing of the for-profit company Human Genome Sciences, established a research arm, the Institute for Genomic Research. Another genome, that of the microorganism *Mycoplasma genitalium*, was completely sequenced at the institute by a team headed by Claire Fraser, Venter's wife.

In 1995, Venter, in collaboration with molecular geneticist Hamilton Smith of Johns Hopkins University, determined the DNA sequence of the entire genome (all the genetic material of an organism) of *Hemophilus influenzae*, a bacterium that causes earaches and meningitis in humans. The achievement marked the first time that the complete sequence of a free-living organism had been deciphered, and it was accomplished in less than a year. The sequence information was expected to aid in the development of a vaccine against the bacterium and illuminate the mechanisms of the infection process.

Meanwhile, a U.S.-government-sponsored effort to se-quence the human genome had been initiated in 1990. Known as the Human Genome Project (HGP), its stated goal was to complete the sequencing in 15 years at a cost of $3 billion by coordinating the work of a number of leading academic research centres around the country, in collaboration with the U.S. Department of Energy and the Wellcome Trust of London. In 1993 the leadership of the project was given to Collins.

Collins, who was born in Staunton, Virginia, had been homeschooled by his mother for much of his young life, and took an early interest in science. He received his bachelor's degree from the University of Virginia in 1970. He earned a PhD in physical chemistry at Yale University in 1974 and a medical doctorate at the University of North Carolina at Chapel Hill in 1977. In 1984 Collins joined the staff of the University of Michigan at Ann Arbor as an assistant pro-fessor. His work at Michigan would earn him the reputation as one of the world's foremost genetics researchers. In 1989 he announced the discovery of the gene that causes cystic fibrosis. The following year a Collins-led team found the gene that causes neurofibromatosis, a genetic disorder that generates the growth of tumours. He also served as a leading researcher in a collaboration of six laboratories that in 1993 uncovered the gene that causes Huntington chorea, a neuro-logical disease.

In 1998 Venter led a new private-sector enterprise, Celera Genomics, to compete with and potentially undermine the publicly funded Human Genome Project. At the heart of the competition was the prospect of gaining control over potential *patents* on the genome sequence, which was considered a pharmaceutical treasure trove, although the legal and financial reasons for the rivalry remained unclear.

The necessity of a government effort came to be questioned when Celera Genomics appeared to be working even faster than the HGP at sequencing DNA. Collins successfully thwarted attempts to merge the public effort with the private endeavours, but in the end the two sides came together, and in June 2000 Venter and Collins jointly announced the completion of the working drafts of the human genome sequence. The breakthrough was hailed as the first step toward helping doctors diagnose, treat, and even prevent thousands of illnesses caused by genetic disorders.

For the next three years, the rough draft sequence was refined, extended, and further analysed, and in April 2003, coinciding with the 50th anniversary of the publication by *Francis Crick* and *James D. Watson* that described the double-helical structure of DNA, the Human Genome Project was declared complete.

STEVEN PINKER (1954–)

Canadian-born American psychologist
at the forefront of cognitive science.

Pinker was born in Montreal. He studied cognitive science at McGill University in Montreal, where he received his BA in 1976. He earned a PhD in experimental psychology at Harvard University in 1979. After stints as an assistant professor at Harvard (1980–81) and Stanford University (1981–2), he joined the Department of Brain and Cognitive Sciences at the Massachusetts Institute of Technology (MIT). In 1989 he was appointed full professor at MIT and became director of the university's Center for Cognitive Neuroscience. In 2003 he returned to Harvard as a professor of psychology.

Pinker's early studies on the linguistic behaviour of children led him to endorse noted linguist Noam Chomsky's assertion that humans possess an innate facility for understanding language. Eventually Pinker concluded that this facility arose as an evolutionary adaptation. He expressed this conclusion in his first popular book, *The Language Instinct*, which became a runaway best-seller and was rated among the top ten books of 1994 by the *New York Times*. The book's best-selling sequel, *How the Mind Works*, earned a nomination for the Pulitzer Prize for general nonfiction. In *How the Mind Works* Pinker expounded a scientific method that he termed "reverse engineering", which involved analysing human behaviour in an effort to understand how the brain developed through the process of evolution. This gave him a way to explain various cognitive phenomena, such as logical thought and three-dimensional vision. In *Words and Rules* (1998) Pinker focused on the human faculty for language, offering an analysis of the cognitive mechanisms that make language possible, and in *The Stuff of Thought: Language as a Window into Human Nature* (2007) he explored what words reveal about brain structure and thought processes.

Pinker's work, while enthusiastically received in some circles, stirred controversy in others. Predictably, there were religious and philosophical objections to his strictly biological approach to the mind, but scientific questions were raised as well. Many scientists felt that the data on natural selection were as yet insufficient to support all of Pinker's claims and that other possible influences on the brain's development existed. Although he conceded that there was much research left to be done, Pinker – along with a considerable number of other experts – remained convinced that he was on the right track.

SIR TIM BERNERS-LEE (1955–)

British computer scientist, generally credited as the inventor of the World Wide Web.

Berners-Lee was born in London. Computing came naturally to him, since both of his parents worked on the Ferranti Mark I, the first commercial computer. After graduating in 1976 from the University of Oxford, Berners-Lee designed computer software for two years at Plessey Telecommunications Ltd, in Poole, Dorset. After this he had several positions in the computer industry, including a period from June to December 1980 as a software engineering consultant at CERN, the European particle physics laboratory in Geneva.

While at CERN, Berners-Lee developed a program for himself, called Enquire, that could store information in files that contained connections ("links") both within and among separate files – a technique that became known as hypertext. After leaving CERN, he returned to England and worked for Image Computer Systems Ltd, in Ferndown, Dorset, where he designed a variety of computer systems. In 1984 he went back to CERN to work on the design of the laboratory's computer network, developing procedures that allowed diverse computers to communicate with one another and researchers to control remote machines. In 1989 Berners-Lee drew up a proposal for creating a global hypertext document system that would make use of the Internet. His goal was to provide researchers with the ability to share their results, techniques, and practices without having to exchange email constantly. Instead, researchers would place such information "online", where their peers could retrieve it at any time, day or night. Berners-Lee wrote the software for the first Web server (the central repository for the files to be shared) and the first Web client, or "browser" (the program to access and display files retrieved from the server), between October 1990 and the summer of 1991. The first "killer application" of the Web at CERN was the laboratory's telephone directory – a mundane beginning for one of the technological wonders of the computer age.

From 1991 to 1993 Berners-Lee evangelized the Web. In 1994 in the United States he established the World Wide Web (W3) Consortium at the Massachusetts Institute of Technology's Laboratory for Computer Science. The consortium, in consultation with others, lends oversight to the Web and the development of standards. In 1999 Berners-Lee became the first holder of the 3Com Founders chair at the Laboratory for Computer Science. In 2004 he was knighted by Queen Elizabeth II. His numerous other honours include the National Academy of Engineering's prestigious Charles Stark Draper Prize, awarded in 2007. Berners-Lee is the author, along with Mark Fischetti, of *Weaving the Web: The Original Design and Ultimate Destiny of the World Wide Web* (2000).

INDEX

Note: Where more than one page number is listed against a heading, page numbers in bold indicate significant treatment of a subject